Toyota
Kaizen
Methods

Six Steps to Improvement

Toyota Kaizen Methods

Six Steps to Improvement

Isao Kato and Art Smalley

 **CRC Press**
Taylor & Francis Group
Boca Raton   London   New York

CRC Press is an imprint of the
Taylor & Francis Group, an **informa** business

A PRODUCTIVITY PRESS BOOK

Productivity Press
Taylor & Francis Group
270 Madison Avenue
New York, NY 10016

© 2011 by Taylor and Francis Group, LLC
Productivity Press is an imprint of Taylor & Francis Group, an Informa business

No claim to original U.S. Government works

Printed in the United States of America on acid-free paper
10 9 8 7 6 5 4 3 2 1

International Standard Book Number: 978-1-4398-3853-2 (Paperback)

Library of Congress Cataloging-in-Publication Data

Kato, Isao.
   Toyota kaizen methods : six steps to improvement / Isao Kato and Art Smalley.
      p. cm.
   Includes bibliographical references and index.
   ISBN 97814338532 (alk. paper)
   1. Organizational effectiveness. 2. Industrial efficiency. 3. Leadership. 4. Toyota Jidosha Kogyo Kabushiki Kaisha. I. Smalley, Art. II. Title.

HD58.9.K374 2011
658.4'013--dc22   2010036076

Visit the Taylor & Francis Web site at
http://www.taylorandfrancis.com

and the Productivity Press Web site at
http://www.productivitypress.com

Contents

Foreword

After spending nearly 20 years at General Motors in various manufacturing positions, I was given the opportunity in the latter part of the 1980s to join Toyota Motor Manufacturing as the general manager of the Power Train Division in the United States. It was during my years at Toyota that I developed a close working relationship with Art Smalley and later Isao Kato.

During my Toyota years, Isao Kato was the principal training manager on various topics worldwide for Toyota Motor Corporation. He personally conducted significant training and development work for us in North America. Isao Kato was the trainer who taught me personally and also hundreds of other team members the basic concepts of standardized work, job instruction, Kaizen, and other topics. Most other Toyota trainers during the past several decades at one time or another have been developed by Isao Kato, and by extension his influence worldwide has been significant. Later, after his retirement from Toyota, Isao Kato helped Art Smalley and me by providing critical training on similar topics at Donnelly Corporation, where we led a successful lean transformation. These courses were critical to our success in both leadership development and implementation of the Donnelly production system.

The unique opportunity this book provides all of us is to take advantage of Isao Kato's nearly 40 years' experience of developing people inside Toyota on topics related to improvement. Art Smalley also has numerous years of experience working directly for Toyota in Japan, as a supplier to Toyota in the United States and as a consultant to Toyota, which is a rare combination of experiences. Their combined experiences help in presenting the various aspects of Kaizen in unique ways. Usually, we receive information from individuals who have merely read books, attended lectures, toured plants, or recited what they think happened. This book originates directly from two of the best Toyota had to offer.

What this book represents goes far beyond the current use of Kaizen as a simple week-long "event" or "blitz" type of activity. What I personally learned from people like Isao Kato and Art Smalley during my Toyota years was that a company's success largely can be attributed to total employee involvement in daily Kaizen. This difference is critical when compared to traditional Western manufacturing companies. Team members in Toyota working with improvement tools, involvement opportunities, and a structured process constantly delivered amazing results that surpassed my expectations.

The engine of the success for Toyota for decades has been how every team member is challenged to conduct waste observations every day, join a participation team to conduct mini ongoing Kaizen events, update the standard from Kaizen, and utilize the new improved method going forward. This book represents the six basic steps required to implement practical Kaizen activities in your organization. Once understood, these steps can be performed and applied throughout the entire company, with the entire team focused on Kaizen.

I recommend for most companies that the skills from this book should first be used to train and implement at the level of team leaders or first-line supervisory individuals. Engineers and managers will benefit as well. However, our training philosophy at Toyota was first to train the main two leadership levels with an expectation that the team leader and the supervisor would immediately begin coaching and implementing the Kaizen methods with their teams. The critical step here is to train and expect your leaders to become coaches and teachers leading their team members to success through the application of the six Kaizen steps. This process, once implemented, builds the knowledge and understanding of waste identification and waste elimination at all levels within the company as leaders are moved and promoted.

This book represents a model for understanding Kaizen inside Toyota and the skills required to analyze basic processes and drive improvement. This is the heart of the Toyota production system, and you can achieve the same degree of success if this Kaizen process is properly deployed within your company.

Russ Scaffede

Former vice president of manufacturing, Toyota Motor Manufacturing
Retired vice president of manufacturing, Toyoda Boshoku America

Chapter 1

Introduction

This workbook is about the topic of Kaizen and focuses on the skills, methods, and analysis techniques practiced inside Toyota Motor Corporation for the past few decades with regard to this topic. Please note that this is not a book about holding Western-style five-day Kaizen events, selecting areas for Kaizen, or detailing best practices for running such workshops. Five-day implementation workshops were in reality quite rare during the development of Toyota's production system and are virtually nonexistent today inside Toyota. In this workbook, we instead focus on the actual training course concepts and methods used by Toyota over the past few decades to *develop employee skill level* with regard to this critical topic. It is our belief that developing employee skill level in topics such as this one and others has always been a core element of Toyota's success.

We drafted the contents of this workbook with several specific goals in mind. One aim is to trace and communicate the main origins of Kaizen since the inception of Toyota Motor Corporation. Another main intent of this workbook is to articulate the basic six-step Kaizen improvement skills pattern that was taught inside Toyota. The steps are similar to other improvement programs in the past as well as problem solving and the scientific method. A third goal is to help practitioners of Kaizen improve their own skill level and confidence with this topic by simplifying it and removing as much of the mystery in the process as possible.

The internal Kaizen skills course at Toyota consisted of a combination of lecture, shop floor observation practice, and some implementation. The chief difference between it and the more common five-day workshop model practiced so widely outside Toyota is the amount of emphasis put on the skills development of the participant. In the Toyota Kaizen skills course, typical participants included first-line supervisors in manufacturing as well as young engineers. The dozen or so participants in the course were each required to learn a six-step pattern for Kaizen, master multiple analysis skills, implement a few simple improvement ideas during the week, and then, on returning to home locations, drive further improvement in areas under their sphere of influence. The roots for all this are made clearer in Chapter 2 on the background of Kaizen in Toyota.

Mainly in this workbook we cover the classroom part of the Kaizen methods course, explaining each step in detail. For some steps of Kaizen (e.g., Chapter 5, "Step 2: Analyze Current Methods"), multiple techniques exist, and we outline those more commonly used. Most of the concepts can be depicted using explanation and simple diagrams. Some of the concepts best require demonstration, and we either attempt to explain the demonstration or provide instructional examples.

In each chapter on Kaizen, we also suggest homework assignments to be conducted independently for further learning. The part of the Kaizen basic skills course that we unfortunately cannot duplicate via workbook is the hands-on observation and implementation practice under the guidance of a skilled veteran. It is our hope in creating this workbook, however, that we can help many improve their own skill level and confidence in Kaizen.

The Kaizen skills concepts explained in this workbook should be of value to you whether you choose to use a five-day workshop model for implementation or some other vehicle for improvement. It is not our intent to prescribe the participation model by which you will drive implementation. Instead, we focus on the simple tools and skills that Toyota taught internally for decades to help individuals succeed in improving processes.

If you take the time to study the concepts detailed here, you will be reviewing the same methods and techniques that were drilled into generations of Toyota supervisors, managers, and engineers. These basic techniques are not the "magic bullet" or "secret ingredient" of lean manufacturing. However, mastery of these timeless techniques will improve your ability to conduct improvement in almost any setting and generate improvement results for your organization. We wish you the best of luck on your improvement journey.

Isao Kato and Art Smalley

Chapter 2

Background of Kaizen in Toyota

2.1 History of Kaizen Methods in Toyota

In this chapter, we briefly review the different influences on the concept of Kaizen inside Toyota and clarify some of its origins. As you will see, there is no simple or single beginning for Kaizen inside Toyota Motor Corporation. Kaizen is not a new word in Japanese, and the notion of improvement was always central to Toyota from the time of the founder Sakichi Toyoda and his son Kiichiro in their initial endeavors related to creating better looms in the early 1900s. In this background section, we briefly highlight some of the influences in Toyota's Kaizen methodology. For those interested just in the actual process and methodology, you can skip ahead directly to the chapters covering the six steps of Kaizen.

The word *Kaizen* in Japanese is written 改善 with two kanji characters meaning "to change" and "for the better." Unfortunately, the origins of the term are not exactly clear in terms of etymology. The word *Kaizen* is Chinese in origin and has roots as far back as the Qing dynastic period in China from 1644 to 1911. The term has always meant improvement, although it was not used exactly in the specific sense we use it today in lean manufacturing, business, or process improvement.

In the early part of the 20th century, the term *Kaizen* gradually started to appear in published Japanese works. However, it was not a word widely used by the general population. Kaizen was mainly used as a technical term in books and did not cross over into the modern spoken vernacular. Starting around the early 20th century, the industrial engineering movement in the United States and other countries made methods-based improvement a priority. Works by Fredrick Taylor Frank and Lillian Gilbreth and others in the field became popular topics. Translations of these books into Japanese no doubt spurred the need for a specific word to mean improvement in this sense, and adaptation of the Chinese characters representing "Kaizen" likely occurred.

Internally at Toyota, the term *rationalization* was often applied to early structural improvements in manufacturing. The term *Kaizen* started to proliferate inside the company in the 1950s and 1960s as an ongoing part of the Toyota Production System (TPS) development. Developing people who could analyze work methods and make improvements (i.e., creativity before capital) was a large priority. Out of such origins, the "Kaizen course" on which this workbook is modeled was born in the Education Department of Toyota and rolled out as training for many decades. Many versions of the Kaizen course exist, and it is not possible to depict all the versions used over the years. However, in a broad sense we can identify some of the main imprints on the development of the concept of Kaizen inside Toyota and the methods used to develop skill and ultimately improve process performance.

2.2 The Toyoda Precepts

Long before Toyota made automobiles, the company was known for making spinning and weaving machines known as looms. Sakichi Toyoda founded several different companies for this purpose and along with his son Kiichiro developed several highly successful automatic looms. These machines, with amazing and innovative design features, are still on display at Toyota's Commemorative Museum for Industry and Technology in Nagoya, Japan. Profits from the loom business and sale of technology are what enabled Toyota to venture into the automotive business in the mid-1930s.

Kiichiro and his adopted brother, Risaburo, codified the main principles[1] of their father as basic tenants of management for the company and named them the Toyoda Precepts. The five main tenants are expressed in Figure 2.1.

Toyoda Precepts

1. Be contributive to the development and welfare of the country by working together, regardless of position, in faithfully fulfilling your duties.

2. Be at the vanguard of the times through endless creativity, inquisitiveness, and a pursuit of improvement.

3. Be practical and avoid frivolity.

4. Be kind and generous; strive to create a warm, homelike atmosphere.

5. Be reverent, and show gratitude for things great and small in thought and deed.

Figure 2.1 Toyoda Precepts.

The second precept, with its emphasis on creativity, inquisitiveness, and pursuit of improvement, highlights the early focus on improvement even during the founding days of the company.

For example, one of Sakichi and Kiichiro Toyoda's great breakthroughs was the development of an automated loom that included automatic stop features in the event of a defect and zero change over time for a shuttle device in 1924. Instead of one person operating one machine, an experienced operator could now operate a bank of 24 to 36 machines. On top of that, defects were caught at the source at the time of thread breakage and not later through inspection. As such, labor productivity, changeover time elimination, and quality were all dramatically improved—talk about the spirit of Kaizen!

Other types of improvements were common as well in the early days of Toyota, including the now-famous concepts of flow, just-in-time production, quality improvements, and creation of more flexible equipment for manufacturing.[2] Arguably, however, most of these improvements were driven by management and were generally "top-down" in nature.

2.3 Training-Within-Industry Job Methods Introduction

Despite its enlightened management tenants and the early emphasis on improvements, Toyota struggled during its initial years as an automotive company and nearly went out of business. In 1950, Toyota faced extreme financial crisis and was forced to reduce its workforce by 2,146 employees. Kiichiro Toyoda stepped down as president of the company to assume responsibility for the situation and was replaced by Taizo Ishida. As a countermeasure to the current situation, management in Toyota worked harder to improve cost competitiveness and embarked on a five-year plan to improve its manufacturing equipment and facilities.[3] It was against this backdrop that machine shop manager Taiichi Ohno started making his radical improvements with his fledgling production system efforts.

A lesser-known fact, however, in the history of Toyota was that in addition to the five-year modernization plan for its facilities, Toyota embarked on a plan during this same time to improve the skill set of its manufacturing leaders. In 1951–1953, Toyota implemented the training-within-industry (TWI) model that had been so successful in the United States during its wartime production effort. TWI outlined five main responsibilities for a leader in manufacturing (Figure 2.2).

With respect to the three skill requirements, TWI established three corresponding 10-hour training courses taught over a 5-day period. Job instruction (JI) taught participants how to train people properly in an effective methodical manner. Job methods (JM) taught participants how to make small improvements in their daily work. Job relations (JR) taught participants how to handle employee-related work problems using a four-step model.

All three of the TWI courses were rolled out in Toyota from 1951 to 1953 by the training department at the company. The JM course is of particular interest as

---

Five Requirements of a Leader

1. Knowledge of work   4. Skill in improving methods

2. Knowledge of responsibilities   5. Skill in working with people

3. Skill in instructing

Figure 2.2 Five requirements of a leader.

this was the first time that Toyota introduced a structured program to its manufacturing leaders (particularly supervisors) for the purpose of developing skill and making small improvements in daily work routines.

Like the other TWI courses, JM follows a simple four-step methodology for making improvements in production. As the original TWI training manual stated, "The development of improvements does not require inventive genius—but it does require the questioning attitude of the supervisor who knows the intimate details. The skill of improving methods can be learned. It provides a way for tremendous savings through more effective use of manpower, machines, and materials."[4] The basic methodology of JM consists of the steps and questions shown in Figure 2.3.

JM analysis was not the starting point for the Toyota Production System. It clearly lacks any linkage to the concepts of flow, just-in-time, built-in quality, Five S, standardized work, visual control, and a host of other techniques for improvement. It also came along several years after Taiichi Ohno started making improvements in his machine shops. However, the TWI JM course significantly was the first structured classroom training course aimed at supervisors and leaders for the purpose of method-based skills development that occurred in Toyota Motor Company.

Unfortunately, the TWI JM course did not last long in Toyota. It was discontinued after a few years for a variety of reasons. As we will see, however, it did have significant lasting influence on the development of Toyota's internal Kaizen course and its contents. In particular, the concept of developing skills for improvement, breaking a job down for the purpose of study, elimination of unnecessary details, use of the 5W 1H (what, why, where, when, who, and how) line of inquiry, and providing a worksheet for participants to complete as an analysis aid were retained and continued.

2.4 "P-Course" Introduction

As mentioned, the JM course inside Toyota enjoyed a short life span. Although it was eventually discontinued, the other TWI courses (JI and JR) continued to be taught for decades. In 1955, the Japan Management Association conducted a seminar in Nagoya near Toyota's facilities; the seminar was called the Production

4 Steps of Job Methods Analysis

Step 1 – Break down the job
1. List all the details of the job exactly as done by the present method.
2. Be sure to include material handling, machine work, and manual work.

Step 2 – Question every detail
1. Use these types of questions:
   a. Why is it necessary?
   b. What is its purpose?
   c. Where should it be done?
   d. When should it be done?
   e. Who is best qualified to do it?
   f. How is the "best way" to do it?

2. Also question the machines, equipment, tools, product design, layout, workplace, safety, and housekeeping.

Step 3 – Develop the new method
1. Eliminate unnecessary details.
2. Combine details when practical.
3. Rearrange for better sequence.
4. Simplify all necessary details.
5. Work out your ideas with others.
6. Write up your proposed new method.

Step 4 – Apply the new method
1. Sell the method to your "boss."
2. Sell the new method to the operators.
3. Get final approval of all concerned on safety, quality, quantity, and cost.
4. Put the new method to work. Use it until a better way is developed.
5. Give credit where credit is due.

Figure 2.3   Four steps of job methods analysis.

Technology Courses, or for short, the "P-courses." A representative of Toyota happened to attend the seminar and reported favorably on its contents to Taiichi Ohno, the vice president of manufacturing.

By 1955, Ohno had completed his initial model lines and accomplished substantial transformation of the engine, transmission, and chassis machine shops inside Toyota. His challenge from top management, however, was how to expand the improvements into other parts of the company and how to develop more people capable of driving the improvements. Contact was made with the Japan Management Association, and an invitation was issued to conduct the P-courses internally at Toyota for the purposes of skills development. The instructor dispatched to Toyota was none other than the now-famous consultant and trainer Shigeo Shingo.

On visiting Toyota for the first time, Shingo, in his own words, was amazed by some of the accomplishments that were already in place. "I started teaching the P-courses in Toyota in 1955. The most astonishing thing I observed was 'one person handling multiple machines.' In the machine shop there were 3,500 machines and 700 operators. On average one person was thus handling 5 machines. The maximum case I heard about was where one person was handling 26 machines and I was very surprised."[5] Toyota had come a long way in Ohno's machine shops in a few years of effort. Still, skill was lacking in the supervisory ranks to sustain and further the developments that were then ongoing.

The P-courses were Toyota's next attempt at developing methods-based improvement skill in manufacturing personnel. This attempt was more successful than the JM course, and a lasting relationship between Toyota and Shigeo Shingo was formed that continued for well over two decades. For example, by 1980, according to Shingo's count, he had taught the P-courses 79 times to approximately 3,000 people inside Toyota.[6]

On average, Shingo taught at Toyota roughly three times per year during his association with the company. All courses were taught as lectures in the classroom with occasional trips to the shop floor for practice observation. The vast majority of courses were several days long; however, a month-long special course was offered every few years as well. The chief participants in the courses were supervisors in manufacturing and young engineers new to the company.

In terms of content, the P-courses in reality were an injection of industrial engineering training into Toyota. Although sometimes referred to collectively as simply the P-course, there were multiple topics taught on different occasions. The chief P-course topics taught at Toyota were as shown in Figure 2.4.

Each topic was taught in a classroom using lecture, examples, and discussion. Assignments were also made, and shop floor observation and practice drills were created. The importance of such education was evident to Taiichi Ohno, as seen by remarks he made to one of us:

> It is important for employees to be able to look at the work they are performing and be able to properly identify waste. Once waste is spotted it is the responsibility of the team to improve the process. The important thing is to teach people to challenge problems and apply the

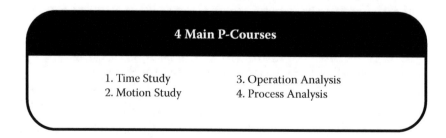

Figure 2.4 Four main P-courses.

```
┌─────────────────────────────────────────────────────────┐
│              5 Main Steps for Improvement                │
├─────────────────────────────────────────────────────────┤
│                                                          │
│    1. Identify the Problem    4. Propose and Evaluate    │
│    2. Establish a Goal           the Method              │
│    3. Identify a Better Means 5. Implement Improvement   │
│                                                          │
└─────────────────────────────────────────────────────────┘
```

Figure 2.5   Shingo's five steps for Improvement.

process of Kaizen. We need to foster the habit in employees of trying to change things for the better.[7]

Unlike JM, however, there was no multistep approach to improvement taught internally as part of the P-course efforts. Later in life, around 1980, Shigeo Shingo published a book in Japanese, *Systematic Thinking for Plant Kaizen*.[8] In the book, he related a basic five-step process for making improvements (Figure 2.5). Although Shigeo Shingo did not teach this specific pattern for improvement at Toyota, one can assume that the sequence identified in the book reflected both his thinking and what he was observing at Toyota.

One of us (Isao Kato) had the unique historical role of being Shigeo Shingo's main point of contact at Toyota for several decades. He organized Shingo's training visits to Toyota and maintained a relationship subsequent to Shingo's departure as consultant trainer at Toyota. The role of Shigeo Shingo at Toyota was highly significant from the point of view of employee training and development. Although Shingo taught training courses at Toyota and did not work directly on the Toyota Production System, he left an indelible mark on young leaders in the company over nearly three decades. The P-courses helped train participant in useful industrial engineering tools and taught people how to view and think about continuous improvement in unique ways. Shingo's role as a trainer and educator was influential and should be recognized as such.

2.5 Development of Toyota's Kaizen Course

The various P-courses continued as training offered in the company until the mid-1980s inside Toyota. Parallel to the P-course offerings, however, was the development of the internal Kaizen course for skills development at Toyota starting in 1968. The Kaizen course continued as a stand-alone item until 1981. From 1981 forward, Kaizen was taught in conjunction with the standardized work course. The basic elements of the Kaizen part of these courses are outlined in detail in the remaining chapters of this workbook. Figure 2.6 is an outline of the original Toyota Kaizen course that started in 1968.

The Toyota Kaizen course obviously was not developed in a vacuum. The course was a logical extension of the Toyoda Precepts, the TWI JM course, and

Original Toyota Kaizen Course

Original Toyota Kaizen Course Curriculum - 1968

Session #	Time Allotment (Minutes)											
	15	30	45	60	75	90	105	120	135	150	165	180
Session 1	Open Session	Kaizen & Group Leaders	The Role of a Manufacturing Leader & Kaizen		The Basic Thinking & Process of Kaizen		Discovery of Improvement Potential		Kaizen Examples & Classroom Exercises		Summary of Session 1	
Session 2	Review	Analyze the Current Methods		Process Analysis Methods	Flow Analysis		Route Analysis		Motion Analysis		Summary of Session 2	
Session 3	Review	Prep for Shop Floor Practice	Current State Analysis of Shop Floor								Summary of Session 3	
Session 4	Review	Barriers to Kaizen	Six Self Questions	Classroom Exercise	New Method Development	Ways to Generate Original Ideas		Make a Kaizen Plan			Summary of Session 4	
Session 5	Review	Shop Floor Practice Generating Original Ideas					Make a Kaizen Plan				Summary of Session 5	
Session 6	Review	Presentations of Analysis and Kaizen Plans					Implementing Kaizen		Evaluate the New Method		Summary of Session 6	

Note: In 1968 in Japan there was still a six day work week resulting in six total sessions. Later it was shortened to five sessions.

Figure 2.6   Toyota original Kaizen course outline.

the P-courses taught by Shigeo Shingo. The Kaizen course borrows elements from each of the preceding training courses and adds unique Toyota elements as well. The developers of the Kaizen course, including one of us, were intimately familiar with the strengths and weaknesses of the preceding programs. Figure 2.7 is a timeline showing the implementation of the contents discussed so far. Timing in relation to the TPS handbooks and suggestion system dates are shown for comparison.

In the next chapter, we describe the six main steps (Figure 2.8) that became the basis for the Toyota Kaizen course. Subsequent chapters explain each step in greater detail as well as provide examples and suggest homework assignments.

In the following chapters, each of these steps is reviewed in more detail to explain the main concepts, purpose, and techniques used in each section.

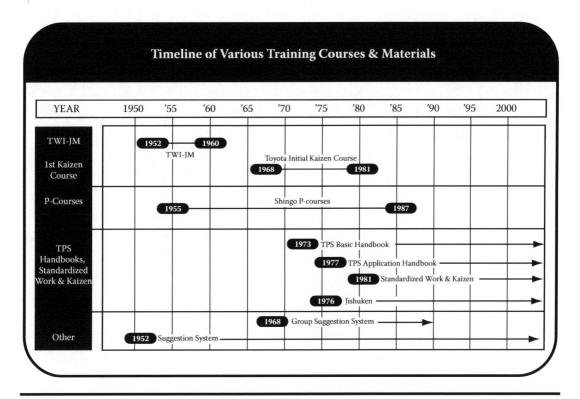

Figure 2.7 Timeline of various training courses.

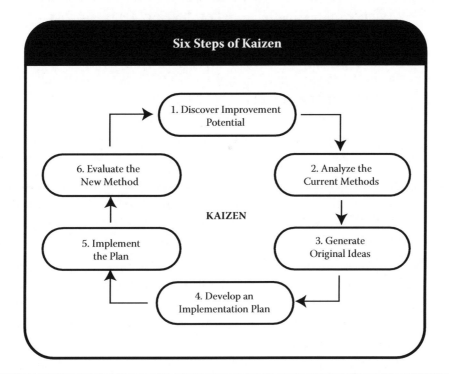

Figure 2.8 Six steps of Kaizen.

2.6 Operations Management Consulting Division and Jishuken Events at Tier One Suppliers

Before diving into the steps of Kaizen, one further topic should be clarified to mitigate some confusion. Within Toyota since 1970, there has existed a special internal improvement group comprised of a select few individuals. This group was initially termed the Production Research Division and was staffed with a handful of Taiichi Ohno's disciples. The group later became known as the Operations Management Consulting Division (OMCD). The role of this group has generally been to help codify and extend thinking on the Toyota Production System, to promote continuous improvement in key locations such as Tier One suppliers, and to develop internal leaders in Toyota by a short—generally two- to three-year—rotational program. One of us (Isao Kato) was a member of the OMCD group for several years.

Historically, the OMCD never played a large role in terms of the number of improvement activities held inside Toyota. The Toyota Production System was largely developed before this group was ever formed. The vast majority of the OMCD work was done either outside the company in Tier One locations well after the Toyota Production System began or in extended workshops involving flow that related back to Toyota. Internal OMCD workshops at Toyota were quite rare, although arguably influential in terms of developmental impact. The OMCD impact in the supply base, however, was large and effective in terms of introducing Toyota style improvement methods in a workshop setting.

OMCD improvement workshops starting in the mid-1970s were called "Jishuken" activities at Tier One suppliers. The term can be loosely translated as meaning "self-study activities." These self-study activities utilized the material developed for the Kaizen curriculum inside Toyota along with other topics but focused more on making change occur in production than on human resources development. The Jishuken activities need to be discussed and put into context because of their later influence on Kaizen workshops overseas. Starting in October 1976, the OMCD office of Toyota initiated structured introduction and implementation of TPS and Kaizen concepts into the Tier One supply base.[9] Jishuken activities were initiated at 17 different companies (Figure 2.9). Each participating company designated a key person in charge of organizing the workshops at that site and promoting the learning points.

The workshops consisted of four basic segments, and these were spread out over time and did not follow a particular time pattern. Participating companies were all located in close proximity to Toyota's headquarters, and multiple sessions over time could be conducted. Participating companies initially each took turns hosting a rotating monthly meeting at which representatives from different companies attended to learn more about Toyota's system and style of improvement.

In concept, the self-study activities consisted of four parts: (1) establishment of a theme for improvement, (2) implementation of Kaizen improvement activities over

Original Members of Toyota Jishuken Activities

No.	Tier 1 Toyota Suppliers	Jishuken Support Person
1	Nippon Denso	Yashio Oya
2	Aishin Seiki	Katsunori Oda
3	Kanto Jidosha	Tetsuo Kondo
4	Kanto Kasei	Hiroyuki Hashimoto
5	Koito Seisakusho	Toshio Horiike
6	Toyoda Gosei	Yoshiki Iwata
7	Aisan Kogyo	Masahiro Kuroyanagi
8	Tokai Rika	Yoshio Okubara
9	Toyoda Boshoku	Yoshihiro Tanba
10	Toyoda Jidoshoki	Shigeru Imada
11	Ishikawa Tekko	Haruhiko Yamada
12	Taiho Kogyo	Chihiro Nakao
13	Central Jidosha	Mr. Iwasaki
14	Takashimaya Nihatsu Kogyo	Tomoichiro Wada
15	Odai Tekko	Ichiro Honno
16	Aoyama Seisakusho	Shyozo Yoshii
17	Jekko	Hiroichi Iwasaki
	Toyota Motor Corporation Support Personnel	Kikuo Suzumura Fujio Cho

Figure 2.9 Original Jishuken members.

a specified period of time—not just five days, (3) repeated trial-and-error style implementation until results were achieved with a presentation of results summary, and (4) final evaluation with critical comments by Toyota's OMCD department.[10]

Some of the Toyota veterans who populated the OMCD office are quite famous, such as the former president and chairman of Toyota Fujio Cho. Other equally talented individuals, such as Kikuo Suzumura, Junichi Yoshikawa, and others, are less well known outside Toyota but no less talented when it comes to the Toyota Production System. Taiichi Ohno never personally led any of the Jishuken workshops at suppliers; however, he acted as an advisor, speaking to the participants a number of times, and provided feedback on implementation.

Sponsoring members at the suppliers learned the basic concepts of TPS by lecture and observation and mostly through the implementation guidance of veterans over an extended period of time. An initial burst of work would typically be followed up and improved on over time, often days or weeks later. Some of these designated key persons who trained under OMCD personnel later retired from their companies in Japan and became somewhat well-known consultants, such as Yoshiki Iwata and Chihiro Nakao of Shingijutsu fame in Japan in the late 1980s and then in the United States in the early 1990s. Although instrumental in helping establish improvement activities in their own Tier One supplier companies, none of these individuals played any role in the development of the Kaizen methodology or overall production system inside Toyota Motor Corporation.

The five-day workshop model of Kaizen frequently practiced in North America and other parts of the world since the latter part of the 1980s is loosely based on the learning points that individuals such as Yoshiki Iwata and Chihiro Nakao and others took from Toyota's Jishuken events. The five-day workshop model adapted and marketed by consultants in North America was an effective package for creating a set period of time that worked for introducing basic lean concepts to overseas companies and implementing some small, yet often dramatic, level of change.

Realistically, it was not always logistically possible to spread out workshops over a larger time frame as had been done in Japan with the OMCD style of Jishuken events around Toyota City. Understandably, this would cause too much downtime for the instructors and potential procrastination on the part of host companies. Thus, in somewhat of a cultural oddity, the Western conception of Kaizen as a large event with a fixed five-day time frame that evolved partly due to marketing and logistical reasons for overseas companies. Toyota did not follow this five-day model or pattern of workshop internally, and that concept is still foreign to the company.

2.7 Summary

Taken together, these different influences combined to create the current types of Kaizen as practiced inside Toyota Motor Corporation for decades and more recently in other parts of the world by many other companies. The actual Toyota Kaizen skills course had its roots in the Toyoda Precepts of Sakichi Toyoda, the TWI JM program, and various industrial engineering concepts as taught by Shigeo Shingo in the P-courses. The six steps of Kaizen as taught by Toyota from the mid-1960s borrowed from all of these sources and others.

The flavors of Kaizen practiced in the United States and other parts of the world also derive from the Jishuken style of events held at Tier One suppliers of Toyota starting in the mid-1970s. The instructors of the Jishuken events of course had working knowledge of Toyota's internal Kaizen course and by extension the P-courses and to a lesser extent the TWI JM course. This knowledge was passed on to their contacts in Tier One Toyota suppliers and then in turn transferred to companies overseas interested in Kaizen.

Regardless of the origins, the style of Kaizen practiced outside Toyota Motor Corporation has chiefly been conducted as an implementation pattern for achieving results and introducing the concepts of Toyota Production System. There is nothing wrong with that intent. Indeed, the purpose of Kaizen is to deliver operational and business-related improvements to the company while developing people. If there is no improvement, then Kaizen itself is an inefficient use of resources. Please note, however, that the purpose of this workbook is to codify the education patterns used inside Toyota to teach the steps, methods, techniques, and thinking patterns used in developing Kaizen skills. These methods and techniques can be learned by anyone, and we believe should be promoted in all types of industries and services interested in improvement.

Kaizen events have been a mixed blessing for many companies that we have visited since the latter part of the 1980s. Some have made tremendous strides in improvement, and others seem to have a hard time sustaining the improvements. In other cases, we have observed employees or work teams reacting defensively toward the topic of Kaizen. It is our hope that by shedding some light on the training practices related to Kaizen skills development, further success can be accomplished by practitioners of this method in all companies and situations.

Notes

1. Toyota Motor Corporation, *Toyota: A History of the First 50 Years* (Dai Nippon, 1988), 39–40. Toyota City, Japan.
2. Toyota Motor Corporation, *Toyota: A History*, 64–73.
3. Toyota Motor Corporation, *Toyota: A History*, 110–113.
4. "Training within Industry Program," *Bureau of Training War Manpower Commission, Management and Skilled Supervision Bulletin* (June 1944): 1.
5. Shigeo Shingo, *Koujyou Kaizen No Taikeiteki shikou* [Systematic Thinking for Plant Kaizen] (Nikan Kogyo Shinbunsha, 1979), 12. Tokyo, Japan.
6. Shingo, *Koujyou Kaizen*, 2–3.
7. Taiichi Ohno, comments to Toyota Education and Training Manager Isao Kato, 1980.
8. Shingo, *Koujyou Kaizen*.
9. Hiroaki Satake, *Toyota Seisan Houshiki no Seisei, Hatten, Henyou* [The Birth, Development, and Transformation of the Toyota Production System] (Toyo Keizai Shinbunshinhousya, 1998), 175. Tokyo, Japan.
10. Satake, *Toyota Seisan Houshiki*, 175–176.

Chapter 3

Introduction to Kaizen in Toyota

3.1 The Importance of Kaizen

Some companies are fortunate, at least for a while, and do not have to focus on rigorous daily execution or improvement-related activities. For example, a company might be the first to market with a hit product, have a technical patent, or possess special technology that grants it a superior advantage over the competition. Most companies are not this lucky, and such competitive advantages do not always hold up in the long run. For the majority of companies in either industry or service, there is a tremendous need to improve or fall by the wayside.

The same scenario has always been true for Toyota Motor Corporation. Toyota started as a tiny player in the automotive industry, making vehicles for the domestic market in Japan. Early company estimates pegged Toyota's productivity as less than one-ninth that of Ford Motor Corporation. Profits margins were also small in Toyota in the early stages of the company. In fact, Toyota almost went bankrupt in 1950 and was forced to release 2,146 employees or about a quarter of the workforce of the company through early retirements, layoffs, or firings.[1]

Companies today must succeed in multiple areas to stay competitive. Product innovation, supply chain efficiency, and internal manufacturing execution are typical areas that require ongoing improvement. The main thrust of this book deals with Kaizen and the improvement methods taught inside Toyota Motor Corporation to employees to improve internal manufacturing operations. Your situation, of course, may be different, and you might start elsewhere out of necessity. The basic concepts and methodology outlined here should help you generate improvement ideas regardless of your respective situation or starting point.

3.2 Key Concepts

There are multiple key concepts involved in Kaizen at Toyota. A complete list would fill an entire book. For starters, here are a few worthy of special mention and explanation that were included as preliminary concepts in the Toyota course. In the following paragraphs, we briefly explain the importance of the role of a leader in manufacturing, different ways to think about increasing production, manufacturing methods and cost, the notion of work versus waste, and the importance of the cost reduction principle.

3.2.1 The Role of a Leader

The first introductory concept pertaining to Kaizen in Toyota involves the expectations placed on any person in a leadership position in the company. By "leader," we generically mean anyone involved in the supervision of employees or anyone who looks after some group of people. Leaders exist at different levels of the company and are a varied lot. In the Toyota Kaizen course, five basic expectations for a leader were referenced for the purpose of general expectation setting (Figure 3.1). The five items are the same ones identified in the training-within-industry (TWI) courses mentioned in Chapter 2.

Specifically, in Toyota's case the three skills mentioned as requirements have all been extremely important over time. The ability to instruct, for example, is critical for training new employees or new people transferring into a department. The act of instruction, when done correctly, can create an atmosphere of trust and respect for the leader. Conversely, when training is skipped or conducted poorly, it leads to a host of problems. The skill of handling people is also critical as disputes and disagreements are going to happen on every team. Having a basic method for dealing with such problems is a great aid in terms of getting teams to cooperate and work together.

The skill in improving methods is also critical and requires proficiency in the two previously mentioned skills of instruction and handling people. Introduction of change almost always sets off a chain reaction of new associated training and dealing with people potentially unhappy about the change. Being a good leader

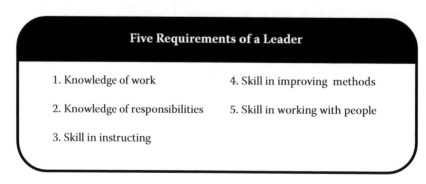

Figure 3.1 Five requirements of a leader.

at a minimum requires some significant capability in all three of these areas, and of course other areas, to succeed.

Improvement in Toyota, however, was not an option. It was an expectation and requirement of all leaders. Some manufacturing leaders might have to work on reducing lead time, while others might have to improve productivity. Still others might have to focus on safety, cost or quality improvements, and so on. The intended message internally was that no one was exempt from the need to improve.

3.2.2 Five Ways to Increase Production

Having established that everyone was expected to improve the work of their team, the internal Kaizen course at Toyota also sought to spell out some specific beliefs about how improvement should be carried out. A typical question posed for discussion was, "How do you increase productivity?" Participants normally responded with typical answers, such as increasing the number of workers, adding machines, working overtime, or working harder. From a sheer numbers point of view, those answers might deliver more units of production, but they do not qualify as true Kaizen. In an ideal case, Kaizen seeks to produce greater quantities of quality product that can be sold using existing manpower, machines, and time constraints. None of the first three typical answers accomplishes that goal, and the fourth one—working harder—is not sustainable or desirable.

In Kaizen, Toyota wanted leaders to be able to separate work quantity input-based improvements (more machines, more time, more people, etc.) from work quality or method-related improvements (e.g., change the nature of the work to be easier and better). In other words, leaders driving Kaizen needed to eliminate waste or unnecessary details in the existing process.

As Figure 3.2 shows, it is possible to make more items by increasing equipment or personnel, but those come at an obvious drawback: increased cost. There are two ways to improve production that do not add cost to the equation, but only one of those ways is desirable from a Kaizen point of view. By improving the quality of their work, teams can in fact produce greater quantities of quality product using existing resources. In modern-day terms, this is of course often referred to as "working smarter" and not "working harder."

The intent of this discussion was to drive home the point to leaders that not only do they have to make improvements as part of their daily job but they also must do it in accordance with the Toyota way of Kaizen. Work details can be analyzed and improved in a variety of ways, as we show in subsequent chapters. The task is initially one of commitment and detailed study. Doing things the right way in Toyota's concept of Kaizen mattered as much as achieving results.

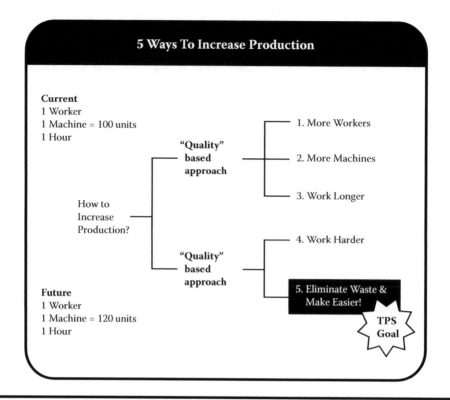

Figure 3.2 Five ways to increase production.

3.2.3 Processing Methods Affect Cost

The third preliminary concept discussed in the Kaizen skills course was the notion that how you performed work eventually affected cost (Figure 3.3). The previous discussion point often drives this point home, but for confirmation the following content was discussed as well.

At the time this course was developed, the intended audience was almost exclusively from the manufacturing ranks. As such, the typical graphic (e.g., Figure 3.3) used was a manufacturing flow sequence that highlighted contrasting styles. Whether or not you are in manufacturing today is of no consequence. The point of the graphic was that there are ways of doing work that involve inefficiency in your current style of operations. That inefficiency might be rework, machine downtime, delays in response times, waiting by personnel, or other problems. It is a leader's task to identify more efficient ways of doing things that involve a better sequence and quality of result.

3.2.4 Work versus Waste (Muda, Mura, Muri)

Sometimes, discussion of the previous concepts caused some employee concern. For example, "I work hard for the company," "I do my best all the time," or "I am very efficient in my day-to-day work routine" are commonly held beliefs. To help reconcile this subjective self-held viewpoint versus reality, Toyota developed the

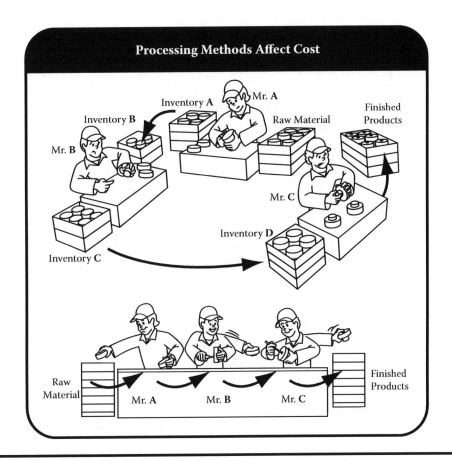

Figure 3.3   Processing methods affect cost.

following concepts over the years of the Kaizen course and Toyota Production System (TPS).

Most people feel they that are very busy at work and sometimes overwhelmed during peak work hours or rush periods. The reality is that most of what people consider "work" is not value added from the customer point of view. Toyota taught leaders to think of work as true value-added operations for the customer, incidental items required in the current state of operations, and pure waste in the operation (Figure 3.4).

True value-added work is a small part of our normal jobs in reality. Customer requirements spell out the form, fit, content, function, and so on of what they desire. The intermediate steps we use to get that end result are usually not specified. A machine such as a lathe, for example, might remove metal to a certain final dimension and surface finish required by the customer. Which exact type of lathe, the tool, the holder, the storage location of materials, the exact program used to make the part, and so on are normally not specified. Only making the required final dimensions and specifications in this case are value added to the customer. The rest of the operation is not entirely value added and can be studied for improvement. In reality, of course, the value-added portion can be analyzed for improvement as well, but that is usually not the initial starting point for Kaizen.

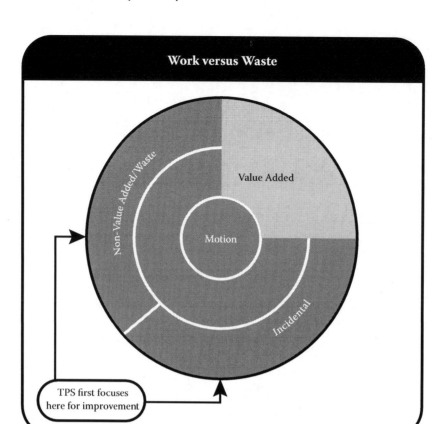

Figure 3.4   Work versus waste.

Incidental waste pertains to work that is required in the current state of the operation that is not valued added but still must be done in the current process. For example, the movement of material is not value added to the customer, but still some minimal amount must be done to get parts from the delivery truck to the process and back again to the shipping dock. Pure waste, on the other hand, is excessively moving materials from one storage location to another location multiple times.

Let us look at a traditional example of an employee assembling some components for shipment to a customer. The same logic holds true for any job. Employee Art Kato of XYZ Corporation walks over to a parts bin and searches for several part numbers. Next, he goes over to a bench and looks through a couple of drawers until he finds a tool he needs. Then, he goes back to his workstation and starts to assemble the items in accordance with customer requirements. In midstream, he realizes he needs one more component to complete the job and goes over to the part rack to obtain that item. After getting this last part, he drives four screws into a housing one by one using a screwdriver. One screw head is accidentally stripped and has to be removed and replaced by another.

It is normal for a person to feel that all of this represents work, and it does represent "effort" on the part of the employee. However, as noted, not all the effort expended is value added from a customer point of view, and in any sequence of operations there is always some area for improvement. In the strictest sense, only the motions that enable the component to reach its final dimension or specification are value added. In the example, multiple forms of waste exist in the form of walking, searching, and repairing, just to name a few examples. As an old manufacturing saying in Kaizen goes, "Only the last quarter turn of the final bolt to specified tightening torque is value added to the customer. Everything else is waste and must be questioned for improvement."

To help leaders and employees see that not all work is value added, Taiichi Ohno coined the terms *Muda*, *Mura*, and *Muri* to explain the concept he was articulating (Figure 3.5). *Muda* is waste, *Mura* is unlevelness, and *Muri* is overburdening the person or process. All three of these phenomena are disruptive to efficient production operations.

Even more specifically, in the mid-1960s Ohno codified seven typical types of waste.

1. Overproduction—Overproduction pertains to producing more than the customer requires or too early. It is both a quantity problem and a timing

Figure 3.5   Muda, Mura, and Muri.

problem. Producing too little or too much, too soon or too late is not acceptable. Overproduction in particular generates excess inventory, the need to handle and store that inventory, and a host of other problems. For this reason, it was referred to as the worst form of waste.

2. Excess inventory—Not all inventories are waste. Try producing without any inventory, for example. Specifically, however, *excess inventory,* defined as more than required to meet customer demand, was considered a waste in Toyota. Too much inventory requires space and manpower hours to move and manage that inventory; also, it carries the risk of obsolescence.

3. Scrap and rework—All forms of scrap, rework, or yield losses during startup and the like were considered a form of waste. Anytime something is not made right the first time, it consumes labor, materials, energy, and more to remake the part. None of these mistakes is considered value added by the customer, and they are considered waste in the process.

4. Wait time—Wait time is another form of waste in the process. For example, employees waiting for parts, materials, instructions, or other such items generate waste in the form of lost time. Once time is lost, it can never be recovered. Ideally, things should always be ready when required with no delays. If not, then that represents wasted time in the process and should be targeted for improvement.

5. Excess conveyance—Excessively conveying materials, as mentioned, is a form of waste. Some conveyance is needed to move between receiving areas to production and back to shipping. Moving parts more than required, however, is not value added. Often, excessive conveyance occurs in conjunction with overproduction and the creation of excess inventory.

6. Excess motion—Similarly, excessive motion by employees can be thought of as waste. A certain amount of motion is, of course, required to complete any task. Double and triple handling of items, multiple trips to get material, or even straining to reach a tool that is improperly located are wasteful. Employees can often feel busy, but not all their motions in reality are efficient or value added to the customer. Teaching employees to realize this and spot it on their own is an important step in the elimination of this type of waste.

7. Overprocessing—The final form of waste in the original list of seven was the waste of overprocessing. This type of waste refers to working items beyond the specifications required by the customer or conducting steps that are not necessary. For example, a part may require a certain tolerance or finish. Continuing to work on the part after it reaches acceptable limits is a form of overprocessing. Putting three coats of paint on when two are sufficient to meet quality standards is another example.

Since the original list of seven wastes was created inside Toyota, many companies have altered the list and added their own forms of waste. Failures to utilize human potential, inefficient systems, wasted energy, and other things are frequent additions to the list. The original list is not perfect and was intended to serve as a way to highlight examples for employees to identify areas for improvement. For parties outside manufacturing, the list requires translation into relevant examples. For instance, waiting for material might instead relate to waiting for documents to arrive or be processed. Scrap or rework might pertain to mistakes in documents or transactions.

3.2.5 Cost Reduction Principle

The final critical concept to discuss as part of the introduction to the Kaizen course is the principle of cost reduction. Kaizen can be conducted for a variety of reasons, including quality, lead time, productivity, safety, and other items. Ultimately, however, in Toyota we were crystal clear as well about the need for cost reduction.

As mentioned, the automotive industry is a highly competitive industry with many complimentary products. Establishing a reputation for quality is critical for any industry. In the long run, companies must also make a profit. A former president of Toyota Motor Corporation, Taizo Ishida, used to remark frequently about the need "to defend your castle by yourself." By this comment, he meant that it was proactive and helpful to take your destiny into your own hands and not leave your personal fate up to others. One of Toyota's methods of embracing this concept was the principle of cost reduction (Figure 3.6).

In the simplest sense, profits are determined for a company by three factors: sales price, cost, and volume. For general discussion, the three elements can be represented by the following equation: Profits = (Sales Price − Cost) × Volume.

Given this simple equation, how can a company earn greater profits? There are only three levers for the equation: increase the sales price, increase the number of units sold, or reduce cost. In general, in competitive industries raising prices is difficult, and customers may simply turn to alternative offerings from competitors. Simply making more product also is no guarantee for making money; the result may just be excess inventory or waste. The only sustainable way to increase profits is to focus on cost reduction.

It is important to indicate that reducing cost does not mean simply cutting costs or jobs. Reducing costs means eliminating waste in any process that does not add value to the customer. Less inventory, fewer defects, less waiting time, and so on all lead to greater productivity of the factors involved in production. This is the true spirit of Kaizen—establish more efficient uses of existing resources by taking out the waste or unnecessary details that do not add value or improve value to the customer. Companies that can accomplish this goal will reduce costs and help improve profits. By emphasizing this formula, Toyota made sure that everyone realized that they had a direct hand in the success of the company.

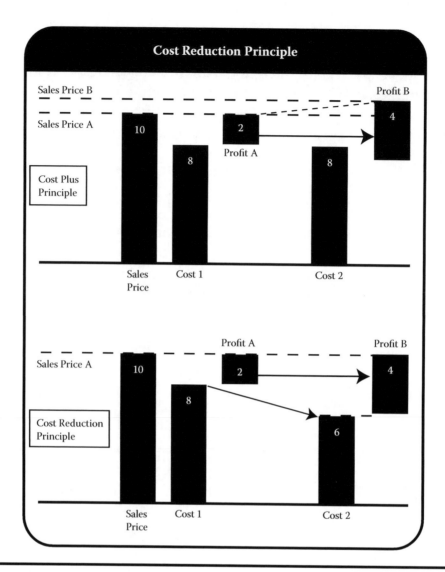

Figure 3.6   Cost reduction principle.

3.3 Basic Pattern for Kaizen

In the remainder of this workbook, we shift gears and move to the concrete steps that Toyota used to teach the process of Kaizen to its internal leaders. In general, there were six main steps of Kaizen (Figure 3.7), and they are similar to other methodologies, such as the scientific method and general problem solving. The big difference is that in Kaizen, as discussed in the remainder of the workbook, there are more degrees of freedom and a greater emphasis on generating original ideas. In general, all improvement methodologies follow the pattern of plan-do-check-act in some basic fashion, and this process is no different.

3.3.1 Step 1: Discover the Improvement Potential

In Step 1, we cover the basic ways that Toyota used to help employees learn to see the waste or improvement potential around them in a more concrete fashion.

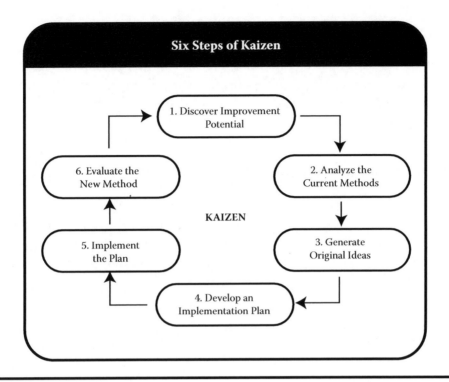

Figure 3.7 Six steps of Kaizen.

We also cover the mindset and attitude that are required for people to be successful in this type of improvement process.

3.3.2 Step 2: Analyze the Current Methods

In Step 2, we review several of the most fundamental ways that Toyota used to teach people how to conduct simple job methods analysis of the current methods. Some of these techniques are old and not unique to Toyota.

3.3.3 Step 3: Generate Original Ideas

In Step 3, we cover some ways to help people get a jump start on generating creative original ideas for improvement. The human mind is the greatest tool that leaders engaged in Kaizen have at their disposal. We review some ways to help get teams moving and thinking in the right direction.

3.3.4 Step 4: Develop an Implementation Plan

In Step 4, we highlight the importance of making a Kaizen plan. Sometimes, the best plan is "just-do-it" type of thinking, and in other cases more coordination and careful thought are involved. We review some important points for teams to consider before implementation.

3.3.5 Step 5: Implement the Plan

In Step 5, we cover the implementation phase of the plan and some key points for consideration here. Often, the best plans do not work as initially intended, are met with initial resistance, and so on. Some general guidelines and key points for implementation are discussed in this step.

3.3.6 Step 6: Evaluate the New Method

In Step 6, we review the importance of verifying results of any Kaizen-related implementation items. In Kaizen, there is no improvement until the results are measured and compared to the previous state. Only results that generate improvement are considered Kaizen, and leaders must make every effort to ensure that a process is indeed improved and not just changed. Remember that Kaizen means "change for the better," not just change for the sake of change. Some common mistakes and points of advice are covered based on our joint experience.

Note

1. Toyota Motor Corporation, *Toyota: A History of the First 50 Years* (Dai Nippon, 1988), 110. Toyota City, Japan.

Chapter 4

Step 1: Discover Improvement Potential

4.1 Introduction

The most difficult part of an improvement process or problem-solving effort is often the first step. In problem solving, the failure to define a problem in proper terms will often derail teams and stop efforts from moving forward. A similar challenge exists in Kaizen. For improvement to occur, individuals involved in the improvement process have to discover the underlying waste and begin to see the improvement potential. The first step of Kaizen is an exercise in helping individuals learn to see different types of waste, inefficiency, problems, and areas for improvement (Figure 4.1).

4.2 Kaizen versus Problem Solving

Before explaining some of the common tools and concepts used to facilitate this first step, let us first take a step back and point out the difference between *problem solving* and *Kaizen*. The two terms are often used interchangeably; however, in Toyota they are distinct concepts. Typically, the definition of *problem solving* centers on the fundamental notion of "gap" or "deviation from standard" (Figure 4.2). A standard may exist for cost, quality, productivity, delivery, safety, or any number of other such categories. When actual measurements of the process deviate from the expected or planned outcome, then a gap exists, and a problem is said to exist.

Problems can always exist in processes depending on how the standard expectation for performance is defined. Sometimes, standards can be set too low,

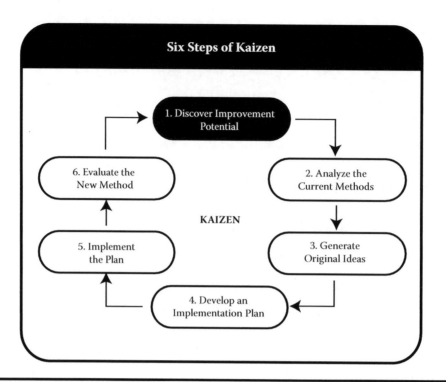

Figure 4.1   Six steps of kaizen (Step 1 focus).

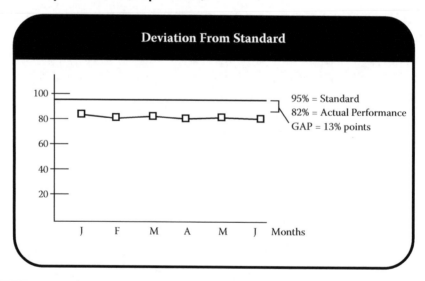

Figure 4.2   Deviation from standard.

or after a period of time the standard is regularly achieved. What then happens when there is no gap from standard? Does this mean that there is no need for improvement? The answer, of course, is "No"; there is always room for improvement. This critical distinction is part of the historical reason for the Toyota Kaizen skills course. Even when a process is operating at standard, we can still expect and drive improvement.

This distinction is minor but critical in terms of difference from problem solving. The thrust of problem solving is usually to get back to standard or an expected level of accomplishment. For example, moving from 95% on-time delivery to 100% on-time delivery is an example of closing a gap or problem solving. Similarly, with quality, productivity, or cost, a gap might exist, and participants study the process to identify a root cause for why the gap exists and then seek to implement countermeasures to close the gap. The notions of "root cause analysis" and "closing the gap to standard" are central to any problem-solving methodology.

What happens when a labor standard of 10 parts per person per hour has been achieved as specified or when on-time delivery is at 100%? In these cases, improvement can still be achieved even though no gap is being closed. Instead of closing a gap to an existing standard, in Kaizen we are looking to achieve a new standard or level of performance (Figure 4.3). In other words, how can we now achieve 12 parts per person per hour, or how can we achieve 100% on-time delivery with a shorter lead time and less inventory?

The difference may seem small to most practitioners of lean, but it has some implications for how improvement is conducted. Problem-solving activities tend to be more quantitative and root-cause oriented and tend to return to existing methods for closing a gap. Kaizen, on the other hand, seeks to establish a new level of performance and thus by definition requires more creativity, degrees of freedom, and a willingness to try new methods. In the end, both respective concepts seek to generate improvements, and the distinction can blur at times. This workbook deals mainly with the latter concept and the notion of improving processes to a new standard of performance.

The one caveat that we offer is that it is often difficult to enact Kaizen on a process that is beset with gaps from standard. Often, a more prudent step is to solve some basic problems in a process to help stabilize its output. Once a process is more stable and predictable, then implementing kaizen on top of that process becomes an easier task. Of course, there is nothing that should stop teams from implementing both problem solving and Kaizen in conjunction if they are

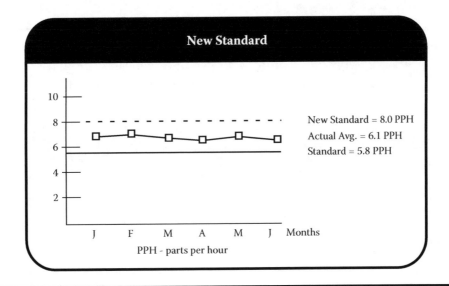

Figure 4.3　New standard.

capable. We leave the topic of problem solving separate from this workbook, however, as it is a topic worthy of discussion in its own right.

Let us assume that you have a process that is ripe for improvement in the sense of the notion that it is achieving standard, but it still has ample room for improvement. There are several characteristics and principles we believe that are critical for getting off on the right foot with Kaizen. In this first step, we cover the importance of proper attitude, analytic thinking, and problem awareness, and then cover some preliminary concepts for highlighting waste and setting goals.

4.3 Kaizen Attitude

Winston Churchill once famously commented, "Attitude is a little thing that makes a big difference." With this mindset Churchill probably would have done well in the area of Kaizen. Much of what you obtain in either problem solving or Kaizen is governed by your attitude on the topic. If you approach the topic with a negative or defensive mindset, you will not accomplish much in the way of improvement. Conversely, if you approach Kaizen with an open mind, a relentless spirit of inquiry, and a willingness to experiment, you will both learn and obtain results.

When approaching the concept of Kaizen, we always suggest that practitioners embrace the mindsets mentioned here. There are other areas that you may add as well depending on your particular circumstances. For starters, here are our top initial suggestions when approaching the topic of Kaizen: (1) Always practice a relentless spirit of inquiry and obtain facts from the actual source or process you are studying. (2) Do not be swayed by preconceived notions or what we jokingly like to call your internal "urban legends." (3) Practice rigorous and thorough observation of the process you are studying. (4) Conduct Kaizen with a calm and rational attitude (Figure 4.4).

Our first piece of advice is to remember that Kaizen is a journey of learning by studying a process and figuring out ways to improve. Unlike problem solving, which often converges to a single root cause or solution, Kaizen encourages more degrees of freedom to obtain new ways of doing things. Of course, simply

Kaizen Attitude

- Get the facts from the source
- Don't be swayed by preconceived notions
- Practice thorough observations (5 Why Thinking)
- Calm attitude (Reason > Emotion)

Figure 4.4 Kaizen attitude.

doing things differently does not constitute Kaizen. The resulting change must be for the better, or the effort is an inefficient use of resources. For individuals to succeed in Kaizen, they must be willing to dig deeply into existing processes and ask the fundament questions of "What exactly are we doing?" and "Why do we do it this way?"

A prerequisite attitude for Kaizen is that individuals must be willing to study any given process and obtain firsthand facts and data about the process. In Japanese, this is called *Genchi Genbutsu* or, directly translated, "go and see the actual location and actual objects." This concept extends to actual tools, drawings, or related bits of the process. Much as a crime scene holds details for detectives, the actual process holds clues for what we are doing, what is difficult, and how it might be done differently. Conversely, in Kaizen we must shun the practice of using old information, secondhand knowledge, or simply practicing group think in a conference room. Rarely will these last activities lead to any real insights about the process or achieve breakthrough thinking.

The second piece of advice regarding Kaizen is not to be swayed by any preconceived notions or opinions about the process. We refer to these preconceived notions as urban legends of the process; for example "that machine doesn't run these parts well," or "that group refuses to do this type of work," or "that process cannot be improved" all fall into this category. Excuses are a convenient reason for not moving forward and are often a tremendous mental barrier when getting started with Kaizen. There may be legitimate reasons why the legend was true once upon a time, or it may simply be nothing more than an incorrect assumption. It is difficult to get out of the gates with Kaizen and study the process with any commitment when these notions are too strong. Make sure all negative opinions are held in check until pertinent facts are gathered, experts are consulted, and new ways of thinking are explored. Otherwise, you are stuck before you even get started.

A third piece of advice is to practice very thorough observations of the process like you never have before. In Kaizen Step 2, we describe some basic ways to study the process in detail. Often, the practice of studying the process in detail will stimulate new questions or insights about how it might be done differently in the future for improved performance.

By thorough observation, we mean two different things. First, be willing to study the process from different angles using different techniques we outline further in this workbook. Changing your angle or viewpoint is sometimes necessary to remove mental obstructions. Second, we urge you to dig deeper with your observations than even before. In this regard, we emphasize, for example, the 5 Why technique that is so often used in problem solving as well as Kaizen (Figure 4.5). Merely dealing with the surface will often stymie your creativity and inhibit understanding of the way things really work.

Our final basic piece of advice is to remain calm and rational during Kaizen. There is a time and place for both reason and emotion in the course of conducting Kaizen. Emotions such as passion and enthusiasm are needed to get started

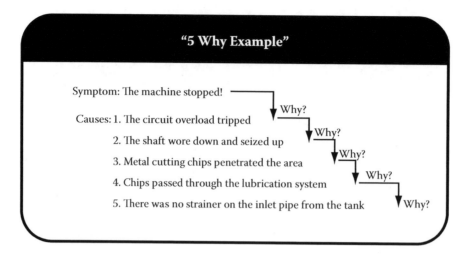

Figure 4.5   5 Why example.

and stay committed when things get rough. The nuts and bolts of Kaizen also require a heavy dose of reasoning and critical thinking skills to be successful. Figure 4.6 outlines the generic steps for the scientific method, problem solving, and Kaizen. Notice the similarities.

Regardless of the method, there is always a detailed method of investigation and commitment to knowledge and improvement. Each method has a slightly different emphasis, and arguably both problem solving and Kaizen are simply derivatives of the older scientific method. The important point we stress in all cases is that it is important to remain calm and rational. Emotional thinking tends to limit the ability to dig deeper, derive insights about cause and effect, or investigate new methods. Use emotion when it is needed but use reason and logic during the investigative phase of Kaizen for expedient results.

Methods Comparison

Scientific Method*	Problem Solving*	Kaizen Steps*
• Make Observations	• Define Problem	• Identify Improvement Potential
• Gather Information	• Analyze Causes	• Analyze Current Methods
• Form Hypothesis	• Set a Goal	• Generate Original Ideas
• Perform Experiment to Test Hypothesis	• Implement Corrective Action Items	• Develop an Implementation Plan
• Analyze Data	• Check Results	• Implement Action Items
• Draw Conclusions & Summarize	• Follow Up/Standardize	• Evaluate Results/Standardize

* Generic patterns. Other versions exist.

Figure 4.6   Methods comparison.

> **Analytic Skills for Kaizen**
>
> 1. Classify and organize
> 2. Quantify the observations
> 3. Specify the details

Figure 4.7   Analytic skills for Kaizen.

4.4 Analytic Skills for Kaizen

In addition to our suggestions regarding attitudes toward Kaizen, we have identified some analytic and quantitative skill sets that are helpful. The three skills in Figure 4.7 apply to all forms of investigation and analysis used in Kaizen. For that matter, the skills are equally important in problem solving.

The first piece of advice (classify and organize) sounds simple but in practice is hard for most individuals and teams to achieve. The overriding tendency is for individuals to present what data are available in whatever format exists. Unfortunately, data presented in this fashion are not useful. For example, let us assume someone suggests that a certain process needs to improve its performance on setup and changeover time. A production department might have a report that a changeover or setup on a machine is currently at 100% of standard. On the surface, this sounds good, but is it? What is standard? How consistently is it being achieved? Is it the largest problem in the process or not?

To answer these initial questions, we need to collect some basic facts and data and organize the information so that initial decisions can be made whether even to pursue this topic. In practice, we urge participants to utilize proper frameworks or analysis tools to frame the situation and put it into proper perspective. Figure 4.8 is an example of a machine and how it can be organized into six convenient categories that are mutually exclusive and collectively exhaustive in principle.

Framing the situation using these six categories for this example should give us a solid framework for thinking about the scope of the opportunity. However, to be more meaningful, the categories also need to be quantified in some manner. Figure 4.9 is an example of how the categories were quantified. For each category, the estimated amount of production lost in terms of pieces was calculated using available data from the past month.

When looked at in this fashion, breakdowns jump out as the largest loss category over the past month on average. Knowing this detail helps point out where to direct further analysis. The danger is that if this sort of quantification is not done, then teams may focus on the wrong improvement topic. For example, working on scrap and rework, while always a valuable area for improvement, in this case is simply not the main priority in terms of gaining more units of production.

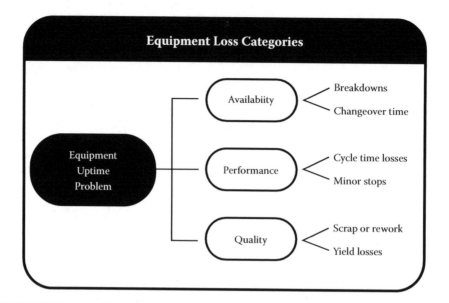

Figure 4.8 Equipment loss categories.

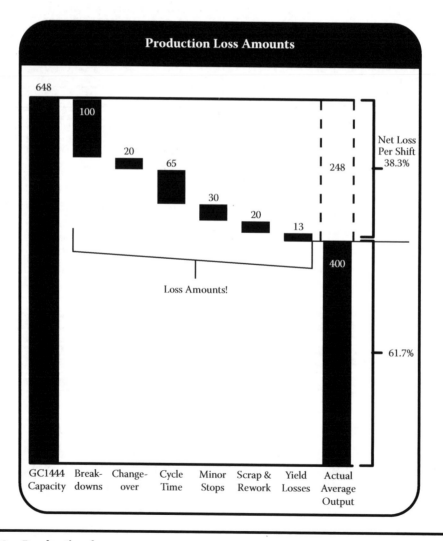

Figure 4.9 Production loss amounts.

After determining the biggest area for improvement, then specific detail in terms of observation and analysis is required (Figure 4.10). This sort of undertaking is case by case and depends on the situation and type of production assets. In this example, let us say that observation identified that a certain proximity switch kept needing adjustment during the shift, and according to the maintenance records it failed every couple of weeks. This single switch malfunction was responsible for 70% of the maintenance downtime during the time studied. Fixing this switch becomes an obvious priority and a way to get a quick win in terms of restoring lost units of production.

These principles of being analytic, quantitative, and specific sound simple, but in reality they are the struggling point for most people. Most of the time, individuals are content to study what data are available rather than to dig and obtain what is required for Kaizen. We suggest extra time and practice in this area to develop your Kaizen skill in this important dimension.

4.5 Opportunity Awareness

Opportunity or problem awareness is another key ingredient in successfully executing Step 1 of the Kaizen process. Analytic skills are part of the mental mindset needed for Kaizen; however, they need to be coupled with an open mind and an ability to be aware of opportunities. As simple as this may sound, we find it difficult in reality to teach and establish. For example, in response to the question, "How are things going with this process today?" a common answer is, "No problem!" In Kaizen, the mindset of "no problem" or "no opportunity" must be carefully avoided.

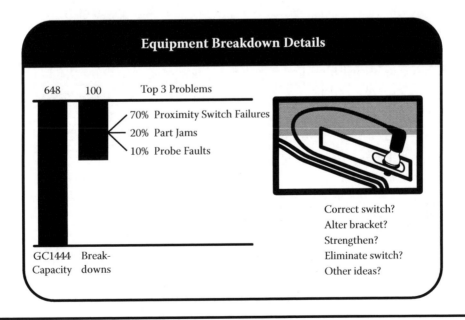

Figure 4.10   Specific detail on breakdown loss.

In reality, problem or opportunity awareness often takes time to develop and a certain level of inquisitiveness as well as persistence. Let us use a historic example from science. Legend has it that Sir Isaac Newton saw an apple fall from a tree, and from there he discovered gravity. There may be a kernel of truth to the story; however, there is likely much more to the story than this anecdote.

In reality, Newton was aware of certain problems in the observable universe. As an undergraduate, he was said to have formulated some rudimentary thoughts on gravitation. In 1664, Newton was visited by Edmund Halley, who discussed planetary motion with him. Considerations regarding this topic led to Newton's formation of his three laws of motion. Application of the laws of motion to Johannes Kepler's laws of orbital motion led directly to Newton's laws of universal gravitation. Only the presence of some force could lead to the observed, curved motion of the planets or an apple falling from a tree, and Newton kept at it until everything was consistent, not only conceptually but also mathematically.[1]

The Isaac Newton example is from science and of course not from industry or service. However, the parallels in terms of thought process are the same. An open inquisitive mind aware of problems (in knowledge or processes) and seeking answers and alternative ways of doing things is an ally in this endeavor. Conversely, a closed mind that cannot spot either problems or opportunities will stymie progress in every case.

4.6 Basic Methods for Uncovering Waste and Identifying Improvement Opportunities

To help teams and individuals get off to a good start in Step 1 of Kaizen, a few different techniques have been useful over the years. Often, these are not needed, and teams can identify a variety of things to work on in their realm of control. When teams or individuals are stuck, however, these methods might help get things moving in the right direction. In Step 2 of Kaizen, we cover some more structured types of analysis.

4.6.1 Compare Performance to Standards

The first simple technique is to compare recent internal performance to standards for the job. In other words, in an objective quantifiable way, seek to ascertain how well things are actually going. Asking opinions and interviewing pertinent parties may have some value but always seek to balance subjective qualitative views with the facts of the current situation.

For example, the type of chart shown in Figure 4.11 might be constructed for thinking about recent performance in an area. In this case we can spot three problems and one area for improvement. A specific problem for example is that

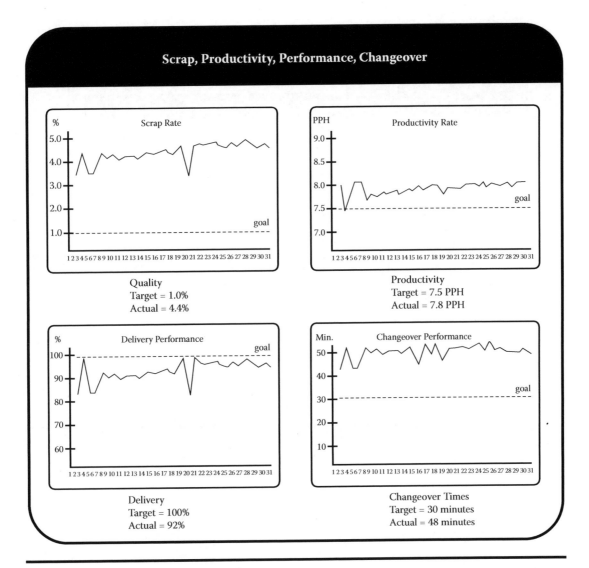

Figure 4.11   Sample performance chart.

the scrap rate through the process is 4.4% while the standard expectation is 1.0% in this case. A gap of 3.4 percentage points exists and could be considered for improvement. Productivity, on the other hand, is meeting standard at 7.5 parts per person per hour and has been achieving this level for some time. Improving work productivity in this area would be a good candidate for improvement activity.

4.6.2 Production Analysis Board

Another starter technique that has been applied over the years in Toyota is the production analysis board. Sometimes, metrics do not always exist at the process level or do not tell the whole story of what is happening. Use of the chart in Figure 4.12 in Toyota forces more detailed hour-by-hour study of the process and can lead to many potential improvement insights. Filling out this chart is often a good starting point for analysis in certain repetitive production situations.

Production Analysis Board

Time Period	Plan	Actual	Difference	Reasons for Difference
8:00–9:00	26	21	−5	• Material late to line −5 min. • One person late
9:15–10:00	31	27	−4	• Tester delay error five times • Broken tool
10:00–11:00	24	22	−2	• Station 3 part jams in chute • Station 5 defects
11:15–12:00	31	27	−4	• Material late to line −5 min.
1:00–2:00	21	21	0	• No delays
2:00–3:00	31	25	−6	• Station 3 part jams in chute
3:15–4:00	26	27	+1	• No delays
4:00–5:00	26	15	−9	• Tester-delay error — maintenance recalibrated machine
Totals	**216**	**185**	**−31**	

Figure 4.12   Production analysis board.

In this example, the expected production rate per hour is compared to actual output. From the chart, a variety of problems and opportunities can be identified. For example, material is late in being delivered to the line at startup and right after lunch. The main tester seems to suffer from probe faults more frequently as the shift continues. One comment notes that material is late to the line and thus costs a few units of production. Another comment notes that parts tend to jam in Station 3 fairly often. Later in the shift, tester errors occurred as well. More detailed observation would of course highlight more opportunities for improvement. An analysis board is a great way to start highlighting some of the opportunities that exist. With some ingenuity you can adapt this board for production, service, or any type of process.

4.6.3 Seven Types of Waste

One way to get participants in the mode for discovery of improvement potential is simply to have them go and look for waste in the process. In the introduction to Kaizen in Chapter 3, we described the seven types of waste and gave several examples for each. Now, it is time to practice observing them in your own operations and see how they manifest in your environment. The concept is what is important for each. The exact type of waste of course will vary by location and industry (Figure 4.13).

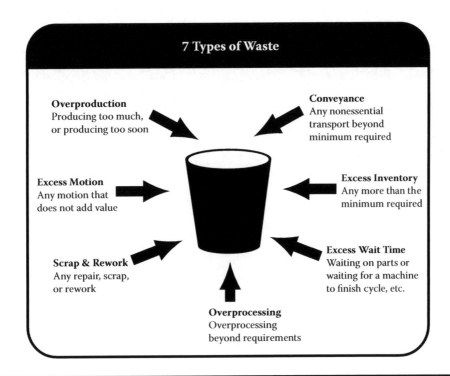

Figure 4.13 Seven types of waste.

- Where is there overproduction?
- Where is there excess inventory?
- Where is there excess conveyance?
- Where is there excess motion?
- Where is there scrap and rework?
- Where is there waiting time?
- Where is there overprocessing?

Have people learn each of the seven types of waste and how to spot them. Remind individuals to be as specific as possible in their observation. "Too much inventory," for example, is too generic a statement. Identify the specific part number and determine how much is on hand in stock. How much should be there? The same advice applies for the other forms of waste as well.

4.6.4 Five S

Often, the simplest solution is the best solution. In many cases when individuals cannot see improvement potential, it is a good idea to start with the basics, such as Five S. This concept applies to five Japanese words starting with the letter S (*seiri, seiton, seiso, seiketsu,* and *shitsuke*). It is possible to translate the terms into English equivalents, but something is usually lost in the process. Figure 4.14 shows the five words in Japanese and an interpretation of what each one truly means.

Five S is about much more than just cleaning. It entails an entire process for improvement in an organizational sense. Five S not only cleans up and organizes

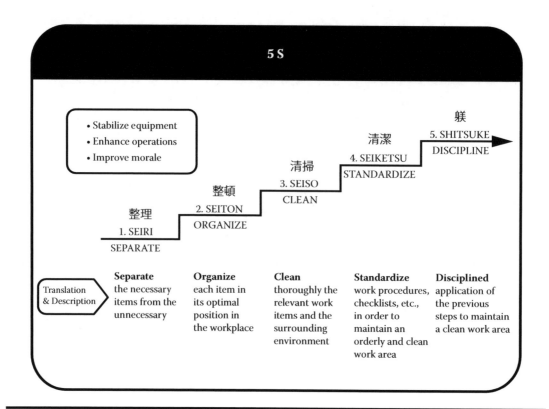

Figure 4.14 Five S.

an area but also usually uncovers problems and opportunities for improvement. For example, machines that leak oil might be discovered. Critical work surfaces that affect dimensions might be contaminated by foreign objects. Work tools or items required to run the process might either be missing or be hard to find. Spending some time working on Five S in an area is a great way to discover some areas for improvement.

4.7 Summary of Step 1: Discover Improvement Potential

In this chapter, Step 1: Discovering Improvement Potential, we covered some of the most critical items for getting started with Kaizen. In particular, it is important to have the right attitude and analytic skills for success in Kaizen. Fortunately, these are learned behaviors, and we can all get better at them over time. Learn to keep an open mind when getting started with Kaizen. Practice classifying observations, make quantitative measurements, and be specific and not just general in these.

In the beginning, a couple of simple techniques might help participants learn to see improvement opportunity more readily. Compare current performance to expected standards for performance. Where are you meeting standard, and where are you lagging? A production analysis board can help you get quantitative and specific when observing a process. For some people, taking a walk and observing the forms of waste in a process is a good technique. Simple Five S-type housekeeping can be a great way to discover some improvement ideas as well.

4.7 Homework Assignment

For this initial Step 1 in Kaizen, a simple homework assignment is highly recommended. If you are working with a group of people, divide into small teams. Of course, this can be one individual as well. Conduct each of the following activities and have everyone bring notes back to the classroom.

- Compare standards to actual performance
- Create a form of a production analysis board
- Identify the seven types of waste in an area
- Implement Five S in an area

Note

1. Dan Berger, http://www.madsci.org/posts/archives/1999–04/924706083.Sh.r.html.

Chapter 5

Step 2: Analyze Current Methods

5.1 Introduction

In this chapter, we present some of the basic techniques available for studying the current methods of various work-related processes. All of these techniques have their roots in industrial engineering, operations research, or similar fields. Some of the techniques have been adapted and modified by various practitioners, including Toyota, over the years. Originally, most of these methods were created for studying manufacturing processes, but with a little imagination and creativity, you can devise ways of using them in most settings, including service operations, health care, and other fields. The concepts of time, motion, work elements, and flow in particular are applicable to just about every situation you will ever face.

Ideally, these analysis methods should be explained first in a small group setting and then tested on actual processes in your own area of responsibility. The Kaizen skills course at Toyota was built on this concept of simple classroom demonstrations and then shop floor actual practice.

In this chapter, we present an overview of several techniques in concept and then outline the general demonstrations used for instructional purposes. We explain each of the analysis forms and present a completed example for discussion. With this level of explanation, we believe you should be able to attempt the analysis methods on your own and gain experience from trial and error. In the appendix section of this workbook, we provide sample forms and add a few more details regarding the specifics of the analysis. Without demonstrating the techniques in person, this is the best explanation that we can provide.

5.2 Basic Analysis Methods

In this section, we cover six basic methods for studying work-related processes. Interested practitioners should be able to devise different ways to make the methods work for their respective situations. Unfortunately, there is no such thing as one perfect analysis technique. Each of the following methods has its own respective strengths and weakness. We suggest that you practice using each of the methods for the purpose of skill development. In reality, the type of process you are studying and your specific improvement goal normally dictate which you will use. Still, it is often useful to study processes from different points of view for waste identification and idea generation. We start with the more fundamental techniques and then work up to some more advanced topics. You may have other forms of analysis that are more suited for your particular situation. We present this list only as a simple starting point for the majority of situations.

5.2.1 Work Analysis

The first basic technique that we cover is that of work analysis. In some way, shape, or form, everyone has conducted this analysis whether they realize it or not. Work analysis is simply the practice of writing down the main work components of a job and then putting the items one by one through a deliberative thought process for improvement. Work analysis can be as simple as informally writing down steps of a job using pencil and paper or the more structured and formal process that we outline here. This technique is applicable to just about any process and is a good one to master for all students of Kaizen.

5.2.1.1 Work Analysis Units

Before embarking on work analysis, we should clarify an important point about scope and detail. There are different levels of detail that you might choose to analyze a job, so we'll present our general guidelines on the topic. For starters, consider the breakdown of work outlined in Figure 5.1.

At the highest level in this example is a job performed by an individual. Often, people identify themselves by the type of work they perform. "I am a welder" or "I assemble widgets" are examples of identifying a job but telling us precious little about the contents of the actual work at hand. Of course, even higher levels of work aggregation than this exist for establishing categories, but for simplicity we start here.

One level down from the job category is what we call the task level of work. For example, a person working in reality performs many smaller tasks during the process of performing an operation. Assembling Parts A and B together and then assembling Parts C and D together to form larger Component A are examples of such an assembly task. These begin to give us some level of detail about the job but not always enough to generate insight for improvement.

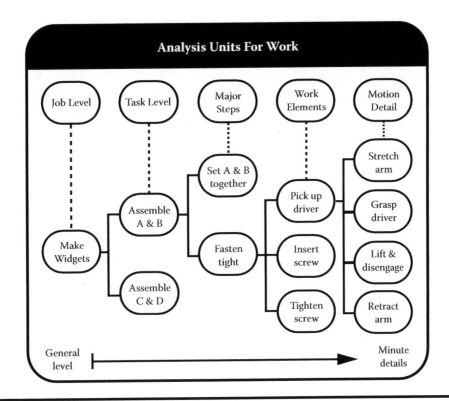

Figure 5.1 Analysis units for work.

One level further down is what we refer to as major steps of the operation. For example, to assemble A and B together, the person must align A and B and then fasten the parts tightly with a screw. At this level, we begin to have some initial detail regarding the specifics of the operation and can generally begin to formulate improvement ideas for consideration.

Below this work level, however, there are still more detailed elemental steps that can be used to describe and further break down work content. For example, fastening a screw involves detailed steps, such as obtaining a screwdriver, inserting a screw, and then fastening the screw tightly as shown in the breakdown in Figure 5.1.

Furthermore, true motion analysis can be performed when needed to go even deeper. The act of obtaining the screwdriver can be broken down into basic motion elements, which are studied further in this chapter. For example, to obtain a screwdriver, the operator in this case has to locate the screwdriver with his or her eyes, extend an arm, grasp the screwdriver, lift it off the table, and carry it to the exact location where it is needed.

These five basic levels are what we collectively call *work analysis units*. The exact level of analysis unit that you use will of course depend on the job being analyzed. As a general rule, we suggest that students of Kaizen first learn to recognize the different levels of analysis that can be performed for work study purposes. Each level has its appropriate purpose, whether it is for general job description reasons or for minute detail observation work.

In general, we suggest that using the middle level of major steps or work element analysis represents a good starting point for most forms of work analysis. This is because at this level in most cases the various forms of wasted motions, inefficiencies, difficult steps, and so on become obvious. Of course, that is not always true, so you will always need to keep in mind the different levels of analysis units.

A few words of caution are in order. If you study work at too macro a level, such as the far left in Figure 5.1, you run the risk of missing improvement opportunities. "Assembling parts," for example, sounds like value-added work to the customer, and in fact it may be in some cases. Analyzing jobs at this level might proceed quickly, but it potentially misses the more detailed waste in the operation.

Conversely, the opposite is true as well. Every step has some waste or inefficiency in it if you look at it in enough detail. Tightening screws, for example, to final torque seems like value-added motion to the customer, and few would question it as a step in an operation. However, you can look at the range of motions required, for example, to tighten the screw (e.g., locating, reaching, grasping, lifting, carrying). Not all of those motions are value added even though they exist in the detailed step of fastening. This level of analysis is extremely useful for highlighting waste and improvement opportunity. Unfortunately, it can also cause problems by being too detailed and time consuming. There may be bigger fish to fry.

The trick in this analysis technique is to pick the right level of detail for analysis depending on the job, the scope or work, and the length of the operation involved. In general, we suggest the middle ground of either major steps or work elements for analysis as a starting point. With a little application practice, you will quickly obtain the insight of whether this is sufficiently detailed for your needs.

5.2.1.2 Training-Within-Industry Job Methods Analysis

One of the first introductions of structured work analysis came to Toyota in the 1950s via the training-within-industry (TWI) job methods course. The primary technique used in the course was simply referred to as the job breakdown method and consisted of a simple form that contained three columns. As Figure 5.2 shows, work details for an operation are presented on the left side of the form in the first column. "Walk to a box of raw materials" and "Pick up five pieces of material" are examples of work content that might be detailed for consideration.

A second column consists of space for the person performing the job methods analysis to write down observation-related notes. The form suggests several categories (i.e., "Reminders," "Tolerances," "Distance," and "Time") for consideration.

Finally, a third column exists on the right for ideas to be presented. The job aid is designed so that improvement ideas and suggestions can be captured one by one as the job is studied. In reality, the ideas might occur during the analysis

Job Methods Breakdown Sheet

OPERATION: _____	PRODUCT: _____	DEPARTMENT: _____
YOUR NAME: _____	OPERATOR'S NAME _____	DATE: _____
DETAILS: List all details for the current method.	NOTES: Reminders - Tolerances - Distance - Time - Etc.	IDEAS: Write them down. Don't trust to memory.

Figure 5.2   Job methods breakdown sheet.

or afterward. A key feature of the form is to capture the ideas and not simply commit them to memory.

On the back side of the job breakdown analysis sheet, the TWI program gave practitioners specific advice for proceeding with analysis. The general pattern recommended was to practice what is known as the 5W 1H (what, why, where, when, who, and how) technique, followed by the principle of ECRS (eliminate, combine, rearrange, simplify) (Figure 5.3). This simple concept is at the heart of job methods and by extension the Kaizen process of Toyota. Why is the described step necessary? What is its purpose? Where should it be done? When should it be done? Who is best qualified to do it? What is the best way to do it? These six simple questions are fundamental for both studying a process and generating ideas and eventually different ways of conducting the job.

More specifically, for what- and why-related questions, the goal is to determine if the steps or details are necessary. If the steps are found to be unnecessary for some reason, then the goal is to eliminate these unnecessary details. In other words, as Taiichi Ohno stressed decades ago, "eliminate the waste in the process" and improve the efficiency.

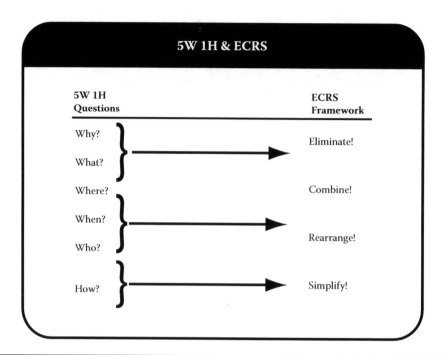

Figure 5.3 5W 1H and ECRS.

Since it is not always possible to *eliminate* details of work, another analysis lens was also placed on the steps. Tactically, for the questions related to where, when, and who, the objective was either to *combine* or to *rearrange* the sequence to improve the work content. First, seek to eliminate and when that is not possible, try to combine or rearrange steps to gain improvements.

The last suggestion in job methods was related to the how question in the 5W 1H method of analysis. For any details that could not be eliminated, combined, or rearranged for improvement, the angle *of simplification* still existed as a last resort. In other words, seek to simplify what remains after having considered elimination, combination, and rearrangement concepts for improvement. If you conduct this form of work analysis in conjunction with the 5W 1H principle for inquiry and the ECRS lens for improvement, suggestions can be identified for virtually any work content you might consider. The assertion is true whether we are talking about steps in an assembly job, steps in taking customer orders, or steps in a quality control check.

For many parties satisfied with implementing rudimentary Kaizen, this form represents a great starting point for almost any job you can identify. All jobs in service or industry consist of processes containing steps, and these steps can be broken down for work content analysis as outlined.

5.2.1.3 Toyota Work Analysis

The TWI job methods course did not last long inside Toyota, as mentioned in previous chapters. However, this concept of work analysis via 5W 1H and ECRS have remained constant inside the Kaizen course until the present. The modifications made by Toyota included emphasizing work analysis units more concretely and altering the form used for analysis and idea generation. Different versions were created over the years. As an example, a fairly common chart is exhibited in Figure 5.4. The intent of the chart is the same as the one used in job methods analysis.

In the appendix section of this workbook, we describe how to fill out the work element analysis form in more detail and provide a blank form. Figure 5.5 is a completed example of a form and a sample task that was studied.

As you can readily see, application of either the TWI job methods form or the sample Toyota version is a great start for beginning Kaizen. An individual or two can study a process and identify work elements as well as generate improvement ideas. A simple team meeting can be held, for example, to discuss the 5W 1H aspect of the form as well as the principle of ECRS. Idea generation will follow naturally in most cases. In the discussion of Step 3 of Kaizen, we also outline some more specific ways to help get started generating original ideas.

The strength of work analysis is its inherent simplicity. The form is easy to use and requires little or no explanation or training beyond defining units of work analysis. The concepts of 5W 1H and ECRS are simple and intuitive to most

Figure 5.4 Work analysis sheet.

Completed Work Analysis Sheet

No.	Work Elements	Safety Distance Dimension Quality Ease	WHY	WHAT	WHERE	WHEN	WHO	HOW	Improvement Ideas	E	C	R	S
1	Obtain material	4 steps away			x				Move closer. Angle tray			x	x
2	Set and organize		x						Eliminate	x			
3	Align edges	Difficult to align		x					Improve guide				x
4	Punch holes	OK							No change				
5	Remove from jig	Difficult						x	File down edges				x
6	Rivet	OK							No change				
7	Inspect	Hard to see						x	Improve lighting				x
8	Stamp date	No longer needed per customer	x						Eliminate	x			
9	Record date	Data field unclear						x	Revise form				x
10	Pack part	4 steps away			x				Move closer			x	x

Figure 5.5 Sample completed work analysis sheet.

people as well. With little training, this document can be used to study just about any type of work. Other adaptations of the worksheet are of course possible if you envision a different format or usage.

Work analysis as presented here does possess certain limitations. The method is neither very analytic nor quantitative in a rigorous sense. It does break down the work elements of a job; however, it does not measure the time associated with each of the elements or measure the cost or difficulty. (Note that time study is discussed as another basic technique for analysis in this chapter.) This method simply takes an existing process and provides a vehicle for breaking it down for detailed examination.

Regardless of its shortcomings, we suggest conducting work analysis as a great way for work teams or individuals to get started with developing their Kaizen skill levels. Few people are uncomfortable with this method, and it is the easiest and fastest way to get started. In industrial cases, this method is also applicable in sales, design, engineering, purchasing, or manufacturing. Those in service-type operations as well as health care practitioners can easily adopt this method with ease.

5.2.2 Motion Analysis

A second basic technique utilized in Kaizen by Toyota is motion analysis. Most students of Kaizen think of "time and motion studies" as the same technique. However, as we highlight, they are in reality different. Motion analysis was introduced to Toyota via the P-courses taught by Shigeo Shingo starting in the mid-1950s. However, motion analysis is several decades older than that and dates to the work of Frank and Lillian Gilbreth, a husband-and-wife team in the early part of the 20th century in the United States.

Most people are somewhat familiar with the story of Frank Gilbreth's work in the study of the motion involved in laying bricks. Gilbreth once famously reduced the number of motions involved in laying bricks from eighteen to four and a half.[1] An article in the *New York Times* in 1911 extolled the virtues of this motion analysis technique, proclaiming that the art of bricklaying had yielded to science for the first time. "If the Pharaoh could only come back now he would rub his eyes at the change made in a trade that had been stationary since he built the pyramids."[2]

Less well known, however, is the fact that Gilbreth and his wife pioneered a basic system called *Therblig* (Gilbreth spelled backward with the t and h reversed) for describing basic motions of workers. The husband-and-wife team also pioneered the use of video capture in motion analysis. Through use of their methods, the Gilbreths helped pursue what they termed the "quest of the one best way" of doing things.[3]

Motion analysis is important to study for several reasons in Kaizen. First and foremost, motion is one of the seven wastes coined by Ohno. That alone makes it an excellent fit for Kaizen in the Toyota Production System. However, there are other benefits to studying the topic, as we will explain. For example, as we will indicate, the study of motion is an excellent way to develop participants' eyes regarding the opportunity for improvement in any process involving human activity. Motion is also a key ingredient for properly understanding other topics for improvement, such as standardized work. Mechanical motion analysis of robots or equipment is of course also possible but beyond the scope of this basic workbook.

Basic motion analysis as taught in the Toyota Kaizen course is most easily understood via introduction of the Therblig symbols. For decades, these symbols were taught to heighten sense of motion and to identify ways to improve. Figure 5.6 represents the basic Therblig symbols and their definitions.

These 18 basic symbols are used to describe most forms of human motion in detail. Items such as walking, however, are not included since this technique mostly applies to stationary jobs involving detailed motion. Demonstration of each of the symbols is the best way to introduce the technique and to quickly memorize them. Then we can use a common example to illustrate the concept. Further detail about the symbols is available online at the Gilbreth Network Web site (http://gilbrethnetwork.tripod.com).

Motion Analysis - Therbligs			
⬯	SEARCH	◖	INSPECT
⬯⬤	FIND	⬤	PRE-POSITION
→	SELECT	⌒	RELEASE LOAD
∩	GRASP	∪	TRANSPORT EMPTY
⌣	TRANSPORT LOADED	ᛁ	REST FOR OVER-COMING FATIGUE
9	POSITION	⌒○	UNAVOIDABLE DELAY
#	ASSEMBLE	⌐○	AVOIDABLE DELAY
U	USE	ᛁ	PLAN
⫝̸	DISASSEMBLE	⌂	HOLD

Figure 5.6 Therblig symbols.

As a classroom example, the following simple demonstration can be used. A person uses his or her right hand to pick up a pencil from a table and then sets it down a few inches away. When most people observe this motion and are asked to explain what just occurred, they simply state something along the lines of "The pencil was picked up and set down" or "The pencil was moved." Those statements are correct, but they are closer to examples of what we used in work analysis to depict the main steps of the operation. The advantage of motion analysis is that we can go much deeper. Figure 5.7 is an example of the Therbligs that might be used to describe the motion involved in the example. As you can see, there is more going on than initially meets the eye. To move the pencil, the following detailed motions occurred as seen by the Therblig symbols:

- The eyes had to locate (search, find, and select) the pencil to be lifted.
- The arm had to be extended (empty transport) forward toward the pencil.
- The pencil had to be secured (grasped) by the thumb and fingers.
- The pencil had to be lifted (disassembled) from the table.
- The pencil had to be moved (transport loaded) over several inches.
- The pencil had to be set down (assembled) on the table.
- The fingers had to let go (release) the pencil.
- Finally, the empty hand returned (transport empty) to the starting location.

Therblig Pencil Example		
LOCATE PENCIL ON TABLE		SEARCH, FIND, & SELECT
EXTEND HAND		TRANSPORT EMPTY
GRASP PENCIL		GRASP
PICK UP PENCIL		DISASSEMBLE
CARRY PENCIL		TRANSPORT
SET PENCIL ON TABLE		ASSEMBLE
RELEASE PENCIL		RELEASE
RETURN HAND TO ORIGINAL POSITION		TRANSPORT EMPTY

Figure 5.7   Therblig pencil example.

Using this description, it takes 10 basic Therbligs to describe the simple motion that just occurred. More complex movements, of course, comprise more Therbligs and often involve the use of the right and the left hand at the same time and use more of the symbols.

Therbligs are an excellent way to describe motion details in a graphical manner. As we mentioned, however, Therblig symbols are also a powerful tool for improving the skill of observation in most people. The technique is often referred to as developing greater "motion awareness" or a "motion mind." Part of the objective of the application of Therbligs is to make students realize just how much wasted motion there is in almost any process involving human work. Figure 5.8 is an example of a sequence of events where a nut is picked up and transfered from one hand to the other, then inspected and set down.

Although the act of "assembly" is thought of as mostly value added, this is not always the case. If you break down a job in your company, you will quickly see how much waste there is even in well-thought-out jobs. If you look closely, most of the Therblig symbols are not value added and simply involve looking, transporting, or holding or imply empty-handed motion. Only a couple of the Therblig symbols actually illustrate valued-added work in the given example. In

Therblig Nut Example

KAIZEN IDEAS	LEFT HAND			RIGHT HAND		KAIZEN IDEAS
	WORK ELEMENT	EXPLAIN	THERBLIG	EXPLAIN	WORK ELEMENT	
	Select parts	Extend hand to parts		Hand waiting		
		Grasp parts		Hand waiting		
		Pick up part		Hand waiting		
		Re-position while carrying		Extend hand		
	Pass to right hand	Release part		Grasp part	Obtain part	
		Retract hand		Re-position while carrying		
		Hand waiting		Inspect and set	Inspect part and set on jig	
		Hand waiting		Release part		
		Hand waiting		Retract hand		

Figure 5.8 Therblig nut example.

addition, most of the work in most operations is done by one hand or the other at a time and does not make use of both hands simultaneously, as would be ideal in many cases. Therbligs can thus highlight waste in terms of efficiency of overall motion.

Figure 5.9 depicts what we mean regarding value-added motion. Therblig symbols can be categorized into three groups: true value-added motion, incidental motion, and wasted motion. Studying Therbligs in detail not only highlights just how much waste exists even in supposed value-adding processes but also provides some hints in general on how to improve.

The clear strength of motion analysis using Therbligs is the ability to focus participant behavior on extremely detailed aspects of work. Even operations that have been improved over the years still demonstrate large potentials for improvement when studied under the specific lens of Therbligs. Once students grasp the concept of Therbligs, then basic aggregate motions (e.g., picking up a pen or assembling parts) are never looked at in the same way. In other words, students learn to see with greater motion awareness.

CATEGORY	THERBLIG SYMBOL	KAIZEN POINT
Analyzing Therbligs		
1. VALUE-ADDED THERBLIGS	# ∪ #	• OPTIMIZE OR SIMPLIFY
2. AUXILIARY THERBLIGS		• ELIMINATE CONVEYANCE • REDUCE DISTANCE • SIMPLIFY • BALANCE MOTION
3. WASTEFUL THERBLIGS		• FIVE S • ELIMINATE THE NEED

Figure 5.9   Analyzing Therbligs.

There are several weaknesses of Therbligs and this particular type of motion study. For starters, it works best on detailed tasks but does not work well when applied to longer jobs. Analysis would simply take too long if hundreds of steps were involved. You can easily wind up looking at individual blades of grass instead of bushes or trees by way of analogy. Second, as critics long ago noted, the method lacks a certain quantitative element, namely, that of time measurement. There is nothing stopping you from adding time to a Therblig style of motion analysis, but the original technique did not include time as a measurement. Indeed, time-and-motion studies are what most people attempt to conduct in many instances today. Last, there are a few instances (e.g., walking) for which the motion is not really captured by Therblig symbols since the method was developed for more stationary jobs.

Whether you use Therblig symbols or a more modern technique such as videotaping to study motion, you should find tremendous opportunity for improvement in any process. The problem is usually one of getting people to see the real details of the process and to realize that much of what is going on is not value added. This sort of analysis trains the mind to spot motion in greater detail and to hone in on various wastes in the process. We recommend analyzing operations in general and then selectively diving down into detail when needed using this type of motion analysis. We call this T-shaped analysis with more general framing at the higher level and then detailed drill downs as required. The goal is not to use the technique everywhere as that would not be practical. The pragmatic path is to focus this powerful improvement method where it is most suited.

5.2.3 Time Study

While most people are probably not familiar with the Therblig style of motion analysis, almost everyone is familiar with the practice of time study. Indeed, most of the time when someone mentions a "time-and-motion study," in reality he or she is mainly conducting a time study with the main work element listed on an analysis form. The Therblig level of detail is not included in the study. In this section, we outline the basics of conducting a time study and contrast it in particular with work method analysis and motion study.

The concept of time study is quite old in manufacturing and predates the Kaizen course at Toyota. Time studies were also common in the loom business of Toyota long before the automotive division was established as a separate company in 1937. The basic concept of a time study is accredited to Fredrick Winslow Taylor and the methods outlined in his work, *The Principles of Scientific Management*.[4] As Frank Gilbreth commented in 1912, "Time study is the art of recording, analyzing, and synthesizing the time of the elements in any operation, usually a manual operation but it has also been extended to mental and machinery operations."[5]

The concept of time is important in Kaizen and the entire Toyota Production System. Time is a major underpinning of improvement efforts in Toyota, whether through the usage of "takt time," "lead time," or "just-in-time" concepts, for example. Time is often an objective, simple way to quantify and measure a process. Along with quality, productivity, and cost, time is at the forefront of metrics evaluating processes. Time can be used to measure duration of events or intervals between events or to help identify sequencing order of concurrent events and so on. Normally, time study in conjunction with Kaizen focuses on either operator motion or machine cycle times. Due to the nature of this workbook, it is not possible to adequately describe all the details of a proper time study; however, we can point out its typical uses, highlight a couple of examples, and review the strengths and weaknesses of the technique.

Filling out a time observation form (Figure 5.10) requires significant instruction and practice. In the Kaizen course, Toyota focused considerably on this basic technique of time study and the practice of using a stopwatch. First, practice was conducted in a classroom setting, then application was made on the shop floor. Since the majority of the participants of the Kaizen course were often not industrial engineers, the basic technique and suggestions for time study were kept simple. The points in Figure 5.11 outline the main concepts for simplicity and practicality.

The practice of time study is unfortunately difficult to explain in text form. However, explaining the basic mechanics, key points, strengths, and weaknesses is still possible. For starters, to explain the basic concept of time study a simple

Time Observation Form

Time Observation

PROCESS							OBSERVER							DATE	
STEP #	Work Element	1	2	3	4	5	6	7	8	9	10	11	12	TASK TIME	REMARKS
	TIME FOR 1 CYCLE													LOWEST REPEATABLE CYCLE TIME	

Figure 5.10 Time observation form.

Tips on Time Study

Steps	Suggestion
Step 1.	Observe work area. Learn the basic cycle and motions.
Step 2.	Write down the work elements.
Step 3.	Measure the total cycle.
Step 4.	Calculate the individual splits.
Step 5.	Find the most repeatable times. Adjust as needed.
Step 6.	Measure and reconfirm any additional items (if needed).
Step 7.	Measure off line work (if needed).

Figure 5.11 Tips on time study.

exercise can be used in the classroom as a demonstration. Further shop floor practice is of course required for mastering the technique.

In the example, a person simply walks to a flip chart and writes down something on the chart, such as "time study." For the sake of practice, the start and stop points for measurement are identified, and a simple time study is conducted via stopwatch. The total cycle time for the demonstration and each individual measurement segment are kept uniform the first several iterations. Then, the demonstrator alters his or her walking pattern, fumbles with the marking pen, and writes time study backward (yduts emit) slowly and deliberately for the last cycle. The result is a cycle that takes a few seconds longer (Figure 5.12).

This example was conducted specifically to drive home certain points for discussion. By design, the instructor would keep certain work elements to a consistent level and vary other ones as measured in the example. The purpose of the variation was to drive conversation and discussion regarding several points. For example, what work element was the longest? Why did it take so long? What work element varied the most? Why did the work element vary? Time reflects motion and thus is an important key to look at for improvement opportunities. Analysis by time does not provide an answer to a problem; however, it provides highly valuable clues regarding where to look.

Time study has certain strengths and weaknesses, as does any operational analysis method. The main strength of time study is that it is quantitative. Time reflects motion and provides a layer of insight beyond writing down work elements or minute motions. Time is also a constant unit of measurement that can be applied as needed to measure manual work, machine cycle times, or auxiliary tasks such as setup and changeover. This is also true whether the focus of the time study is writing a purchase order, taking an order over the phone, or entering an engineering change order into a system. Time is also a great way

Time Study Exercise			
	Cycle 1	Cycle 2	Cycle 3
1. Stand up	2"	2"	2"
2. Go to flip chart	8"	7"	8"
3. Pick up marker	2"	2"	6"
4. Write text	5"	6"	11"
5. Put down marker	2"	2"	2"
6. Return to chair	8"	8"	8"
7. Sitting/Waiting	3"	3"	3"
Total Time	30"	30"	40"

Figure 5.12 Time Study Exercise.

to compare before and after situations and highlight the level of improvement accomplished by any change in method that has been implemented.

A weakness of time study is that it is often more difficult to conduct than work analysis, for example, and often it intimidates people. Individuals not familiar with time study may be reluctant to use the technique, and employees are often sensitive to the practice as well. Toyota's goal with time study was to simplify the tool and put the skill set for measurement into work teams for the purpose of measuring their own work and indentifying improvements. If members of the production team could conduct the time study under the direction of the team leader, this was usually more effective than its being measured by external parties.

Time study is also difficult to conduct when the measurement points are poorly defined for the work elements or the job has not been adequately defined up front. Time study loses some of its meaning when the operation is not performed the same way each time during the study. For these instances, it is highly useful to enlist the aid of someone with an industrial engineering background to provide help in difficult cases.

Time study was a major tool in both the P-courses of Shingo and the Kaizen course of Toyota. Despite its initial difficulty, the skill is valuable in every manufacturing shop for Toyota. The concept of time as a measurement and comparison technique is also vital for the Toyota Production System as a whole. With no time analysis, there can be no just-in-time style of production or standardized work which relies upon takt time. Time dimensions along with quality and cost are extremely important factors for the success of any system.

5.2.4 Standardized Work

What do you get when you combine the techniques of time, motion, and work analysis into one form? In Toyota's case, the combination resulted in a unique analysis method known as standardized work. Along with the kanban of just-in-time fame, standardized work is perhaps the most famous tool in the Toyota Production System arsenal. Standardized work is unique to Toyota and originated within the company over several decades. The concept builds on time, work, and motion analysis to provide an efficient solution for balancing work with respect to demand. In reality, it is not possible to do full justice to the topic of standardized work in this workbook. However, we'll touch on the gist of the technique, introduce one form of standardized work, and discuss how it is tightly linked to the concept of Kaizen.

Standardized work uses a particular document in Toyota, and in its purest narrow sense, the application is somewhat limited. Nevertheless, the concept is important to understand correctly and can be powerful when applied or adapted to fit different settings. Standardized work uses a document (Figure 5.13) that centers on human motion and combines the elements of a job into the most efficient sequence currently possible without waste to achieve the most efficient

Standardized Work Chart

Standard Work Chart			PLANT:		PRODUCT:		
			AREA:		Op. ____ of ____		
			PROCESS:		Pg. ____ of ____		
DATE:	BY:	APPROVED BY:	SHIFTS:				
			VOLUME:				

NUMBER	MAJOR STEPS	Man. time	Auto Time	Wait Time	Walk Time	◯ WORKING SEQUENCE — WALKING ······ RETURN TO START	SAFETY ✚	SWIP ⬛	QUALITY ◇QC

Figure 5.13 Standardized work chart.

level of production. Since the early 1980s, standardized work and Kaizen have been taught together inside Toyota in many instances.

In Toyota's case, there were some fundament prerequisites for correctly applying standardized work. First, the work performed had to be cyclic and repeated throughout the day. Second, there needed to be relatively little difference in work content in products being handled. Third, the quality of the incoming parts and materials needed to be high. Finally, there needed to be minimal equipment downtime. If these preconditions could not be met, then first either more improvement work needed to be performed or a different type of analysis might be more suited to the task of standardizing the job.

Standardized work and the steps of Kaizen are important for helping improve standardization, quality, and productivity. In particular, standardized work was conducted to utilize labor and equipment in the most effective manner especially in areas where human–machine combination possibilities existed. A work cell with 10 stations and three employees highlights a sample situation for consideration. Can three employees provide enough labor to complete the job? What about when demand changes? How many stations should each employee attempt to cover? What improvement possibilities exist?

Standardized work is Toyota's method to help answer these questions. Fundamentally, this analysis technique requires some skill set in work analysis as well as time and motion study. Some important prerequisites exist that are not possible in all settings. Standardized work is used in Toyota on manufacturing lines producing discrete parts with similar amounts of work content. Equipment and quality problems in the line have to be minimal, or they will just highlight the disruptions in the line and potential for improvement. When pure standardized work does not "fit" for whatever reason, we suggest that you instead rely on good job instruction methods first and then layer on Kaizen analysis methods, such as the previous types outlined in this chapter.

When you implement standardized work, a *takt time* (i.e., pace rate of demand) based on customer demand must first be established for the area in question. That number can then be used to identify the extent of work that one individual can manage. For example, if demand is 20,000 units for a given product family this month and there are 22 planned working days with two shifts each with 7.5 hours of available working time (after breaks and lunches), you would obtain a 59.4-second takt time (Figure 5.14).

How many workers are needed in the manufacturing area at this rate? The answer depends on the total work content in the area and its relation to demand. The more work content, the greater the number of workers that will be required. The less overall work content, the fewer workers that are needed. Let us assume that 178.2 seconds of total work content exists to make the product from start to finish based on detailed observations of the process. The total cycle time divided by the part takt time tells us how many resources are needed. In this case, with

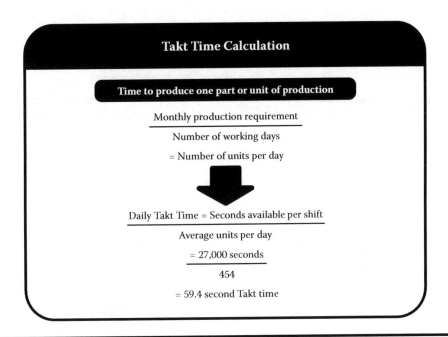

Figure 5.14   Takt time calculation.

a 59.4-second takt time we can thus determine that three operators per shift are needed to complete production since 178.2 ÷ 59.4 = 3.0.

In reality, the math is never this neat and simple. However, the concept is what we are trying to explain. Given the current work content, the most efficient solution is to have three workers performing a job, each with exactly 59.4 seconds of work, no more or no less. More work represents the burden of "muri" on the worker, while less work represents lost opportunity or wasted time. Different line rates based on takt time thus require different staffing levels. Balancing each worker's individual cycle time of work to the demand rate or takt time is a difficult task and yet one that Toyota attempts every month in production via standardized work analysis (Figure 5.15).

The next question for one of these workers is how much work content should be cover given the current takt time. For example, should the worker be expected do Steps 1 to 10 or Steps 1 to 15? In reality, to complete standardized work, several forms often must be filled out in detail. These forms are known internally as the process capacity sheet, the standardized work combination table, and, the most well-known, the standardized work chart. Each serves as an input into the creation of the next document. In this workbook, we only present the final form (the standardized work chart) and link it to analysis for Kaizen. Figure 5.16 is a sample completed form for one operator, for example, working to a takt time of 44.5 seconds in this example. If you observe closely, elements of work, time, and high-level motion are all displayed in some form on the worksheet.

Standardized work links heavily into discussions of Kaizen inside Toyota for several fundamental reasons. For starters, the document neatly summarizes the main elements, work sequence, and times required for each element. If you need to improve a few seconds, then looking at the wait times, walk times, or other problems in the operation is a good place to start.

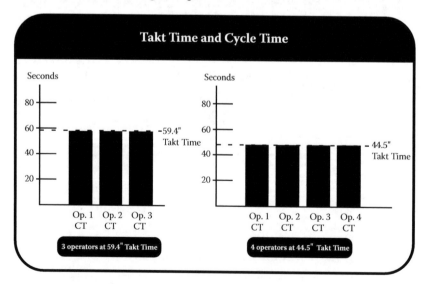

Figure 5.15 Takt time and cycle time chart.

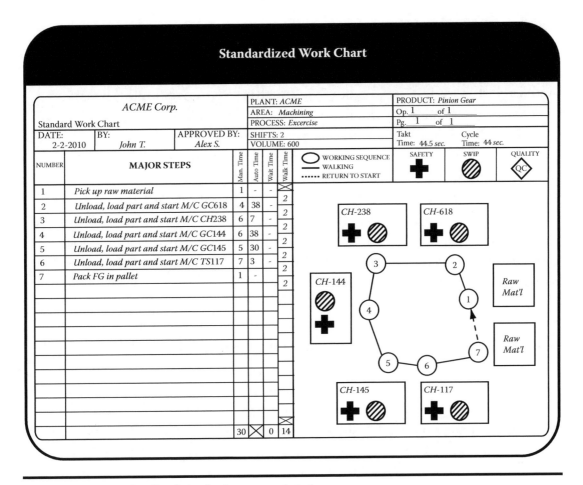

Figure 5.16   Standardized work chart completed.

Second, the number of required work steps for an operation is rarely a nice round number as depicted in a previous example. For example, 3.2 workers might be required to complete a job. However, there is no such thing as 0.2 of a person. The work requires four people in this case unless something changes. Teams are given the tasks of improvement of work flow, elimination of wasted motion and delays, and so on until the task can be completed with three instead of four people. It is important to note that working harder is not the goal here. Elimination of non-value-added activities or waste in the process is targeted and removed to enable the improvement. Standardized work is a great jump start to this analysis process.

Third, standardized work is reviewed and changed monthly in the company since production volumes change and are updated monthly. This change in demand triggers a change in takt time (either faster or slower), and this effect further alters staffing levels (higher or lower) in every department each month. This balance of work around the company is coordinated, and cross-trained workers are moved based upon need and interest level in skill diversification. This avoids trapping the same amount of labor in an area when volumes fall and increases the amount of labor appropriately when demand rises.

Finally, standardized work is used for improvement even when volumes remain constant over time. No one is exempt from the goal of improvement inside Toyota. A work team of 15 people might be expected to perform the same level of work in the upcoming year with one less person, for example. The goal is not to stack up workload, overburdening the worker (i.e., muri). The goal is the elimination of unnecessary work details, wasteful motions, or other non-value-added activities to achieve true productivity improvements, in other words, Kaizen.

The strength of standardized work is the fact that it became the de facto tool for improvement in all operations in Toyota involving human-and-machine combination situations. Instead of having to rely on the three separate forms of work element analysis, time study, and motion analysis, Toyota created a way to merge the essence of all three methods into one analysis tool. Production teams once trained essentially became their own industrial engineer via the use of this form. Standardized work also functions as a quick way for the supervisor to see if employees are keeping pace on the job, following the prescribed sequence of motions, and finding what exactly was varying in the event of certain problems.

There are several difficulties involved in the implementation of standardized work. For starters, it required a 10-hour course in Toyota spread over five days to learn all the associated details of the analysis. Mastering standardized work takes training time and then follow-up commitment by both the trainer and trainee. Second, pure standardized work is often difficult to apply. In most companies, the preconditions (cyclical motion, limited equipment downtime, or lack of quality problems) often cannot be met. The three elements of takt time, work sequence, and standard work in process also might require rethinking of existing product families and equipment layout. Work content variation of great than 15% generally leads to difficulty in implementation. Third, to get the full benefit of standardized work it has to be reviewed and altered at some regular interval (monthly in the case of Toyota). Toyota changes assembly line rates (takt time) in accordance with demand changes and then moves personnel around accordingly. Annual productivity improvement expectations of 5–8% were also placed on teams as well.

Regardless of the difficulty, Toyota has used standardized work as an important analysis tool in conjunction with Kaizen for several decades. Assembly teams in particular in Toyota found the tool to be critical in analyzing current methods of operations and achieving mandated productivity improvements. Along with kanban, the concept of standardized work is the most famous and widely heralded tool in the Toyota Production System. If you do not feel that you can implement standardized work, then it is acceptable to start with proper job instruction training methods and to combine this with some of the previous forms of analysis outlined in this chapter for improvement.

5.2.5 Machine Loss Analysis

The analysis tools covered so far arguably focus on the human aspect of production and ignore the dimensions of material flow and machine work. In the case of Toyota, a disproportionate number of employees in production historically resided in assembly-oriented areas. As such, it is normal to include more tools for analyzing these types of operations. Toyota has a large number of asset-intensive shops as well that conduct Kaizen (Figure 5.17).

Original Kaizen-type activities actually originated in the machine shops of Toyota and not in the final assembly shops, as is often mistakenly assumed. Ohno was an engine machine shop manager and started conducting his human–machine combination experiments as early as 1947, with one person operating two machines, and in 1949, one person was operating three or four machines.[5] By 1955, an operator was on average handling five machines in the engine plant, with one amazing case of a single person handling 26 machines.[6]

Making improvements in a machine-intensive area is different from making improvements in an assembly area. In assembly areas, the value-added part of the work is performed by a person. In machining areas, the value-added part of the work is performed by a machine and tool to form a part. The concepts and tools

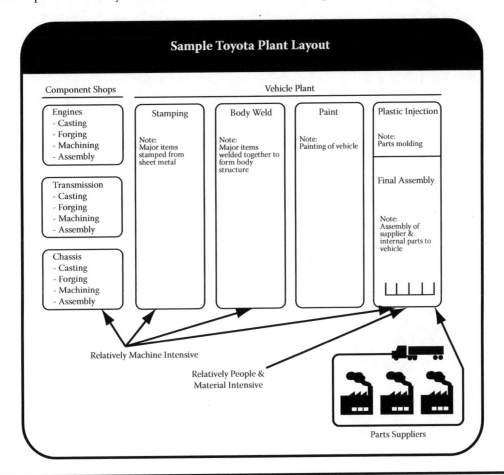

Figure 5.17   Sample Toyota plant layout.

for analysis were all regularly used inside Toyota but generally were not taught as part of the standard Kaizen course. In the hope of helping interested parties with methods for analyzing machines and material flow, we include some more useful concepts in the remainder of this chapter. There are no standard classroom demonstrations for the following methods; however, explanation of the concepts should suffice for participants to understand the basic concepts.

Machine analysis can greatly depend on the type of machine investigated and the production environment. For example, a paint booth is different from a welding machine, which is again different from a machine tool or a piece of semiconductor equipment. The framework in Figure 5.18 can be used in general cases to highlight different areas for improvement.

Students of Total Production Maintenance (TPM) literature will recognize these various losses as the elements that make up the overall equipment efficiency (OEE) metric. In reality, Toyota Motor Corporation never used the OEE metric internally as a regular reporting tool. Toyota suppliers such as Denso and other companies did make regular use of it, however, in some facilities. Internally, Toyota used the six elements that make up the OEE metric as a method for spot analysis in equipment-intensive operations, and in this capacity the analysis tool should be useful for many different parties. With respect to each of the six major losses, we outline some suggestions for analyzing improvement potential.

First, a word of caution is in order. A critically important factor to remember about using Kaizen on equipment is the importance of avoiding overproduction. Merely making more parts on a machine does not constitute an improvement in all cases. Only parts required by customer demand are necessary items to produce. Any production over this amount constitutes overproduction and typically results in inventory. Remember in all of the following categories to be guided by customer demand first and not merely produce to maximum machine capacity.

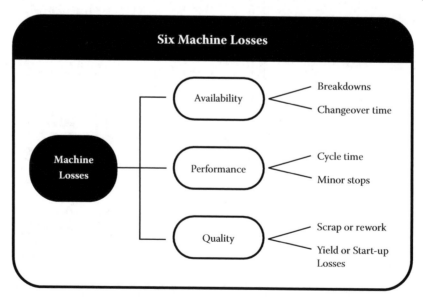

Figure 5.18 Six types of equipment losses.

5.2.5.1 Equipment Breakdowns

There are six main mechanical losses that affect equipment the first of which is typically equipment breakdown. In general, *equipment breakdowns* refers to any unplanned downtime on a machine typically due to mechanical or electrical failure. This loss can be measured over a short period of time involving observation, interview, and data collection. When a proper maintenance history database exists, longer-term data can be used to measure average downtime on a process over a longer period of time.

Reducing equipment downtime on a machine is similar to improving quality problems. Often, the first step is to obtain data on which machine is breaking down the most or which units on a machine most frequently break. Organizing the data into a Pareto chart (Figure 5.19) and Ishikawa-type cause-and-effect (fishbone) diagram is helpful. Once organized, the specific details can be considered for improvement angles.

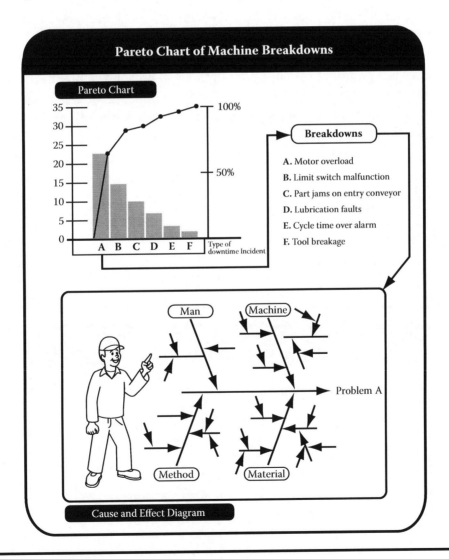

Figure 5.19   Pareto chart of machine breakdowns.

Fixing mechanical breakdowns often requires relevant expertise from either the maintenance or engineering department. It is extremely important to get beyond the superficial details and observe what exactly is occurring. Toyota has relied on its famous 5 Why technique for probing into root causes. Indeed, the most widely quoted 5 Why example is a machine breakdown instance that occurred inside Toyota's machine shops many years ago (Figure 5.20).

As you can see in the example in Figure 5.20, the problem reduced eventually to a countermeasure of attaching a strainer to an inlet pipe on a tank to prevent small metal shavings from entering a lubrication system. Often, this is treated as an ultimate root cause and solution. However, in the spirit of Kaizen we argue that it is neither. For example, further levels are possible for deeper "why" inquiry, and different solutions might work as well. Why did metal shavings or cutting chips enter the tank in the first place? Why did the shavings exit the machine? Was the tank poorly guarded? Was the guarding on the machine insufficient? Was the coolant spray that normally deflects the shavings away to a designated spot inside the machine not working properly? Was the fluid flow or pressure insufficient?

Different countermeasures are possible beyond the strainer on the inlet pipe in this case. Better guarding on either the tank or the machine might be a solution. Also, deflecting the chips differently inside the machine might make sense as well. Different tools, feeds, and speeds or other methods might make a difference in terms of cutting chip formation and how those chips flow away inside the machine (e.g., smaller cutting chips). In Kaizen, we would suggest studying all of these angles for potential solutions. In the next chapter, we discuss the importance of different thinking routines for generating original ideas.

5.2.5.2 Equipment Changeover

A second common type of machine loss is that of lost production time due to changing tools or dies in equipment. Stamping presses and die changeover time

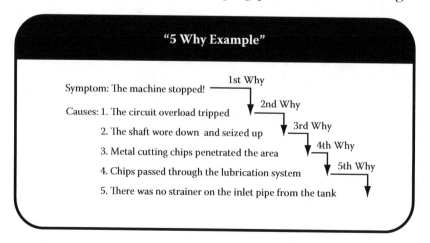

Figure 5.20 5 Why example of machine breakdown.

are the most famous example in this category, but it applies to many other types of equipment as well. The process of changing over a stamping die or other tool may be complex; however, the techniques for analysis are mostly fundamental concepts we have covered.

Entire books have been written about setup reduction, so we will not elaborate on the total process. In its simplest form, reducing losses due to changeover of tools consists of identifying the major steps, then time studying each step and stripping out the elements of waste from the process. Furthermore, great care is taken to distinguish between work that must be prepared before the machine stops (external work) and work that can be done only once the machine is actually stopped (internal work). Moving as much work to the external category from internal work as well as reducing the non-value-added steps in each part of the process can greatly reduce the time forfeited due to this type of machine loss.

The worksheet in Figure 5.21 is highly useful in studying setup and changeover time. If you look closely at the worksheet, the form basically identifies the major steps of the job and the elements of time for each step. For each step of the process, potential problems, improvement points, and countermeasures are established. Repeated application of this concept along with mechanical improvements is what enabled Toyota to reduce changeover time in its stamping

Setup Reduction Worksheet

LINE NAME:		Setup Improvement Analysis						
PART NAME:								
PROCESS NAME:	APPROVED BY:			PART NUMBER				
NUMBER	**Main Set Up Work Elements**	TIME STUDY			CATEGORY		PROBLEM POINT	COUNTERMEASURE
		START	END	TOTAL	INT.	EXT.		
1.								
2.								
3.								
4.								
5.								
6.								
7.								
8.								
9.								
10.								
11.								

Figure 5.21   Setup reduction analysis form.

departments from several hours to a companywide average in 1962 of 15 minutes.[7] By 1973, this level was down to an average of 3 minutes per machine.

5.2.5.3 Equipment Cycle Time

Another type of loss that can afflict machines is related to the mechanical cycle time of the process itself. As we discussed, the concept of time study is stereotypically associated with human work, yet the concept applies to all machine-related work as well. Often, a machine is purchased at a designed cycle time for an operation. The machine may work at this cycle time for many years. Ultimately, however, components wear, maintenance work occurs, and employees change settings in electronic controls and programs that control the machine. In some instances, machines speed up, and in other cases they actually perform more slowly over time.

As a potential improvement area, we suggest time studying machines for multiple reasons. Machine cycle times can degrade over time or be slowed by well-intending parties. Unless cycle times are checked occasionally, accidental loss of production time can occur. A loss of 5 seconds per cycle when multiplied across 1,000 cycles per day can become a large number. Mechanical components can be repaired or replaced, actuators can be upgraded, and programs can often be restored to original settings. Of course, all of this must be carefully investigated in the context of quality and safety.

In addition, even when machines have not slowed over time, there still exists plenty of opportunity to improve. Just because a machine was designed at a certain cycle time does not mean that is the only speed at which it can operate. Often, there is dead time within a machine cycle that can be analyzed for improvement when the situation (e.g., demand now exceeds the capacity of the machine) warrants this type of study. Let us consider the hypothetical cycle and times outlined in Figure 5.22.

In this case, assume that the designed cycle time is 76 seconds and has not suffered any degradation in speed. Demand changes by 20% and requires that more parts be produced off this machine, which suffers from no significant machine losses. Adding equipment is an expensive proposition, and adding time or people adds cost as well. In these cases, the 76 seconds of machine cycle time should be considered for improvement along with other angles. Notice that in this case not all of the 76 seconds are truly value-added machine work. In fact, in this example only 38 of the 76 seconds involve actual physical processing of the component. The other 38 seconds are tied up in small steps to prepare the part for processing. These 38 seconds in particular should be looked at for improvement potential. Seconds can be shaved off incrementally for improvement. in many locations

In advanced cases, even the 38 seconds of value-added cycle time can be studied for improvement as well. Those cases are beyond the scope of this book but should be studied by the relevant mechanical engineers and experts associated

Machine Cycle Time Study

1. Automatic doors open	2"
2. Remove part from machine (or auto eject, etc.)	2"
3. Load next part	2"
4. Clamp part/Coolant on	3"
5. Table index	3"
6. Grinding wheel on (or tool trotates, etc.)	4"
7. Rapid feed advance	4"
8. Air cut	2"
9. Rough cut	(18")
10. Dwell	2"
11. Finish cut	(20")
12. Air cut	2"
13. Rapid feed retract	4"
14. Table/Coolant off/Air blow	4"
15. Unclamp part	2"
16. Automatic door open – Repeat cycle	76"

Only 50% of the machine cycle time is value added...

Figure 5.22 Machine cycle time example.

with tooling, quality, and other aspects of the machine. Optimal equipment settings, tooling conditions, and other aspects of the machine are all fair game for improvement study. For simplicity, we suggest as a starting point identifying the mechanical work elements and time studying each major segment of machine motion.

5.2.5.4 Minor Stops

A fourth category of machine losses is designated as "minor stops" in Toyota. In Japanese, these minor stops are called *chokkotei*. Large equipment breakdowns are typically what catch the attention of management and stand out in the memory of repair employees. However, often machines suffer a variety of minor stops, part jams, sensor confirmation problems, or other slight malfunctions. Senior equipment operators learn over the years how to deal with these symptoms by clearing jams or adjusting sensors on their own. Unfortunately, this does not get at the root cause of the problem, and the situation persists over time. These instances also frequently present safety risks as well.

As with cycle time losses, these small mechanical losses can add up to a large total at the end of the day. Twenty small problems each requiring an average of 2 minutes to fix, for example, can rob production of 40 valuable minutes. Sophisticated machine-tracking systems may identify this type of loss, but in other cases the losses often go undetected. Simple observation of machines and interview of machine operators can often highlight these minor stops for analysis regarding the root cause.

As part of a Kaizen workshop or simply as a matter of daily management, a good practice is to watch machines cycle continuously on a periodic basis. The act of detailed observation with a probing mindset can often highlight the typical minor stops that affect a given type of machine. Once the types of problems are identified, a small check sheet can be used to identify the frequency of the problems if needed. However, keep in mind that the focus needs to be on correction of the cause of the minor stops afflicting the machine and not merely data collection for the sake of data collection.

For example, assume a switch is repeatedly causing problems on a piece of equipment during production time (Figure 5.23). In both Kaizen and problem-solving mode, the cause of the fault is the focus of the investigation. Different consideration points and correct actions might exist to remedy the problem. Of course, regardless of whether problem-solving and root cause analysis techniques are applied or original ideas are generated to solve the issue, the focus remains on elimination of the minor stoppages.

5.2.5.5 Scrap or Rework

Scrap and rework are typical examples of losses in all types of production (Figure 5.24). Often, the time lost in production due to scrap and rework is not so significant; however, the dollar amounts are significant. Scrap and rework problems are more typically analyzed from a problem-solving point of view, and

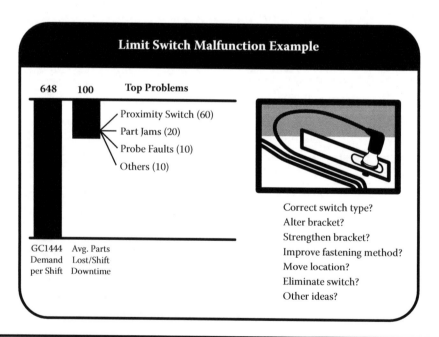

Figure 5.23   Limit switch malfunction problem.

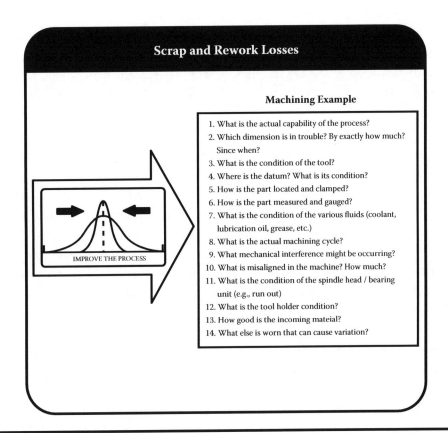

Scrap and Rework Losses

Machining Example

1. What is the actual capability of the process?
2. Which dimension is in trouble? By exactly how much? Since when?
3. What is the condition of the tool?
4. Where is the datum? What is its condition?
5. How is the part located and clamped?
6. How is the part measured and gauged?
7. What is the condition of the various fluids (coolant, lubrication oil, grease, etc.)
8. What is the actual machining cycle?
9. What mechanical interference might be occurring?
10. What is misaligned in the machine? How much?
11. What is the condition of the spindle head / bearing unit (e.g., run out)
12. What is the tool holder condition?
13. How good is the incoming mateial?
14. What else is worn that can cause variation?

IMPROVE THE PROCESS

Figure 5.24 Scrap and rework losses.

we do not suggest altering this line of thinking. However, when solving quality problems, keep a fresh eye on the problem and strive to consider different ways to solve it.

For any given root cause, multiple countermeasure opportunities may exist. We encourage you to keep an open mind during the course of reducing machine losses due to either scrap or rework.

5.2.5.6 Startup or Yield Losses

Startup and yield losses are the final category for consideration in basic analysis of machine losses. Often, when equipment starts up at the beginning of a shift, it is prone to startup problems or yield losses. For example, a machine may run slowly until it warms up. Or, the first several parts off a machine might be more prone to defects until the material reaches a certain stage of processing. Either instance can eat away at the capacity of the machine, and they could be ripe opportunities for study.

In summary, the six losses are a useful spot analysis tool when studying machine-related operations. No machine is perfect, and investigating the six types of losses can often open up improvement angles for consideration. In addition, other topics, such as energy consumption and tooling, might be investigated. Kaizen has stereotypically been applied to manual processes in many companies. We urge practitioners of Kaizen also to look at machine-related cases. Often, improving capital productivity is a tremendous improvement lever in processing. Keep these six basic losses in mind in conjunction with the seven wastes, and you will go a long way in improving any type of operations.

5.2.6 Material Flow Analysis

The sixth and final analysis technique we introduce is known as material-and-information-flow analysis inside Toyota. A popular version of this topic widely practiced outside Toyota recently is called value stream mapping. For all intents and purposes, the terms can be used interchangeably and relate to the same concept. Early forms of the technique exist in industrial engineering as well, as we mention. As with standardized work, full justice cannot be done to the topic of material flow analysis in this limited space. However, we highlight its basic concept and suggest its most effective uses.

The concept of lead time and reducing the time it takes from receipt of customer order until delivery of product and ultimately receipt of payment is an extremely important concept in the Toyota Production System (Figure 5.25). For this reason, the topic of lead time is frequently studied and considered an appropriate topic for Kaizen.

Industrial engineers for many years expressed production flow as existing in one of the following areas: operation, transportation, inspection, and either delay or storage.

Simple process flow charts have been used in operations for decades. Allan Mogenson, who is known as the father of work simplification, stated the following in 1932 regarding process flow charts: "In order to achieve measurement, tools are needed and the most important of these is the process chart. Once a process chart has been drawn up, common sense is all that is needed to improve efficiency and better the process being examined. The process chart then, is

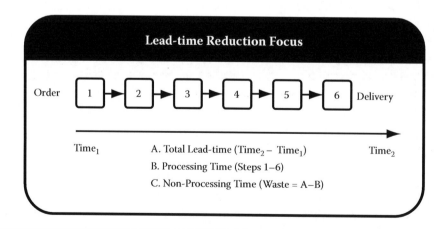

Figure 5.25 Lead-time reduction emphasis in the Toyota Production System.

the lifeblood of work simplification. It is an irreplaceable tool. It is a guide and stimulant. It takes time to properly utilize but there is absolutely no doubt that it works."[8]

Frank Gilbreth of motion analysis fame also is also credited with introducing flow charts as a structured way for documenting process flow in a presentation to the American Society for Mechanical Engineers in 1921.[9] The basic process flow chart symbols referenced in Figure 5.26 were used in conjunction with the Flow Process Chart in Industrial Engineering. Figure 5.27 is an example of a generic order flow process in a 1944 Production Handbook.[10]

This document was used to track the flow of a mail order in a structured fashion and highlight the different problems that might occur in production along the way. The symbols were mapped to represent the flow of the part or process under observation, and to the right side of the form observations were noted. In total, 12 operational steps were recorded, with 4 transportation moves, 3 inspections, and 5 delays. In the legend in the upper left hand corner the 5W 1H (Who, What, Where, When, Why, & How) was listed for questioning purposes. On the right side of the form under "possibilities" and earlier alternative form of the ECRS (Eliminate, Combine, Rearrange, Simplify) framework was used to identify potential areas for improvement. This version uses eliminate, combine, sequence, place, person, and improvement.

Traditional Process Analysis Symbols

Symbol	Step	Meaning
◯	Operation	Alters the shape or other characteristics of the material or product
▽	Storage	An accumulation of materials, parts, or product
D	Delay	Any unscheduled accumulation of materials, parts, or product
☐	Volume Inspection	Measurement of amounts of materials, parts, or products for comparison with specified amounts
◇	Quality	Testing and visual inspection of materials, parts, or products for comparison with quality standards
⇨	Transportation	Changes the location of material, parts, or product

Figure 5.26 Traditional process analysis symbols.

Flow Process Chart

ANALYSIS WHYS					
WHAT?	WHERE?	WHEN?	WHO?	HOW?	

Question Each Detail

Flow Process Chart

No. 1

No. 1 of 1

Job — Special Orders in General Office

□ Man or ☑ Material — The Order Form

Chart Begins — At receptionist's desk

Chart Ends — In mail chute

Charted by — H.F.G.   Date

SUMMARY

	Present		Proposed		Difference	
	No.	Time	No.	Time	No.	Time
○ Operation	12					
▷ Transportation	4					
□ Inspection	3					
D Delay	5					
▽ Storage						
Distance Traveled	140 Ft.		Ft.		Ft.	

POSSIBILITIES

Details of {PRESENT / PROPOSED} Method	Operation / Transport / Inspection / Delay / Storage	Distance in feet	Quantity	Time	Eliminate	Combine	Sequence	Place	Person	Improve	NOTES
1. Waited in box at reception	○▷□D▽										
2. Picked up by confid. clerk	○▷□D▽									✓	Use wire basket
3. Taken to desk at A	○▷□D▽	30'					✓		✓		To files instead — shorter distance
4. Examined (for information)	○▷□D▽						✓				At files
5. Waited (procure info.)	○▷□D▽				✓						Not necessary if taken to files
6. Prices written on order	●▷□D▽						✓				At files
7. Taken to post. clerk at B	○▷□D▽	40'							✓		Shorter distance
8. Placed in desk tray	○▷□D▽										
9. Waited for clerk	○▷□D▽										
10. Picked up	○▷□D▽										
11. Examined (for infromation)	○▷□D▽										
12. Prices added (machine)	○▷□D▽										
13. Total written on order	●▷□D▽										
14. Waited (clerk gets ledger)	○▷□D▽										
15. Total transferred to ledger	○▷□D▽				✓						Taken directly to mail chute
16. Placed in special out box	○▷□D▽				✓						Not necessary
17. Waited for routing clerk	○▷□D▽				✓						Not necessary
18. Picked up	○▷□D▽	40'			✓						Not necessary
19. Taken to desk (C)	○▷□D▽						✓	✓			By B
20. Examined (determine route)	○▷□D▽				✓						Not necessary — save cost of envelope
21. Placed in envelope	○▷□D▽				✓						
22. Addressed to proper dept.	●▷□D▽										Have B route & drop in mail chute
23. Taken to receptionist desk	○▷□D▽	30'						✓	✓		By B — shorter distance
24. Placed in mail chute	○▷□D▽										

Figure 5.27   Sample work flow process chart.

This form is not widely used anymore; however, it still represents an effective way to map a series of steps, classify the items into categories, and specify the details. Each step can then become the focus of further analysis for improvement potential. Production-related work as well as office-related tasks and the like can make use of this fundamental analysis technique.

In Toyota, this type of format gave way to an internally developed document used for what is known as material-and-information-flow analysis. The previous form was highly useful in many instances, but it lacked any linkage to the element of time required to complete the tasks from start to finish (e.g., lead time). As an adaptation, the Toyota version of flow analysis instead looked more closely at the time it took for a product to move from raw materials to finished goods (Figure 5.28). In addition to highlighting the process flow of the product, equal attention was given to the flow of information and time required. Different symbols were also developed and added to represent inventory, types of information flow, and scheduling systems.

The bottom of the material-and-information-flow analysis form highlighted the time component of the equation. Processing time was compared to nonprocessing time for eye-opening purposes. Typically, it might take days or weeks

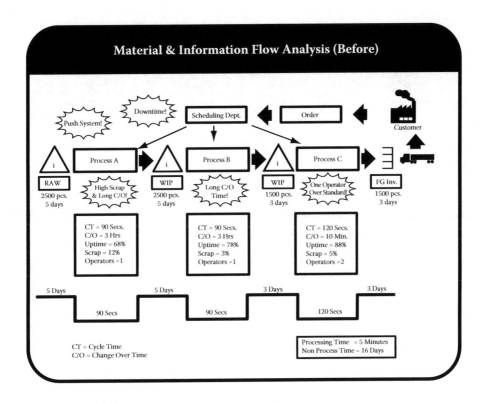

Figure 5.28 Toyota material and information flow analysis before.

for items to flow from raw material to finished goods, yet the actual processing time was measured in minutes. Key points to consider in a diagram of this type include: Where is there too much inventory? Why are delays occurring? Where are we pushing product instead of pulling production? What type of signals are used for scheduling and conveyance? How and where can production be leveled more effectively? What are the chief system inhibitors? The items can then become the focus of improvement activities to aid the flow of the overall system (Figure 5.29).

The goal and chief strength of the Toyota style of flow analysis is the focus on reducing the lead time of the production system measured. In particular, this style of analysis strongly highlights the wastes of overproduction, unnecessary conveyance, and inventory in the Toyota Production System. Hence, this technique is an important tool since it covers what many have called the most sinful of all wastes: overproduction.

In terms of weakness, it can be argued that the material flow analysis tool is high level and does not provide adequately detailed information. It lacks drill down into types of problems such as quality, downtime, labor, or machine productivity. However, in fairness, that was never the design intent of the tool, and there is no reason that these analyses cannot be done in conjunction as needed.

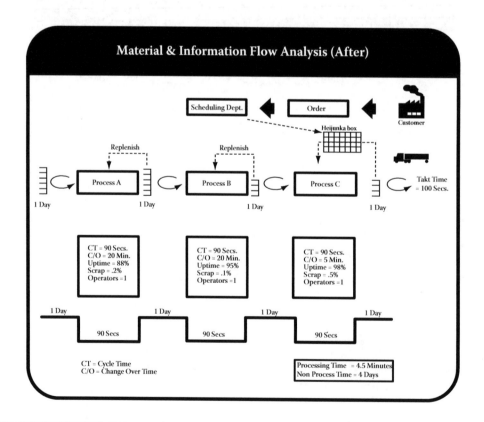

Figure 5.29 Toyota material and information flow analysis after.

As a general point of view, we suggest using this type of flow analysis as the 10,000-foot level framing mechanism for a product family or series of operation. The material and information flow can highlight the breadth and scope of high-level problems. Other analysis techniques discussed can then be used for drilling down into the details of work elements, time, motion, or machines as needed.

5.3 Summary

In this chapter, we covered some of the most basic techniques available for studying the current methods of any process. Almost all of these techniques have their roots in industrial engineering or related fields. Some of the items have been adapted by Toyota or other various practitioners over the years in creative ways. There is no single analysis technique that will work all of the time. Selection of the right tool for the right situation is part of the Kaizen skills development process.

Other methods exist for specifically analyzing quality or cost, for example, and we encourage you to utilize other analysis techniques familiar to you as well. In the appendix section of this workbook, we reproduce the main forms used in analysis and outline the sample steps for completion of each. Time and practice are the only ways to get better at each, so please begin as soon as you see an opportunity to practice the various techniques.

5.4 Homework Assignment

This chapter represents the most critical step in terms of developing your Kaizen skills. There are six main steps of Kaizen; however, this chapter on Step 2, which analyzes current methods, is arguably the most important. We suggest practicing each of the techniques outlined here to develop better understanding of each tool and to develop skill in application. Work elements, motion analysis, time study, standardized work, machine losses, and material flow analysis represent fundamental ways to begin studying a process and identifying improvement ideas. Learn to use each, as well as its strengths and weaknesses, and you will be well on your way to becoming proficient in Kaizen.

Notes

1. David Ferfuson, (The Gilbreth Foundation). http://gilbrethnetwork.tripod.com/bio.html.
2. "Bricklaying Yields to Science for the First Time," *New York Times*, April 2, 1911.
3. http://gilbrethnetwork.tripod.com/bio.html
4. Fredrick W. Taylor, The Principles of Scientific Management (Mineola, NY: Dover Publications, Inc., 1998). This Dover edition, first published in 1998, is an unabridged republication of the volume published by Harper & Brothers, New York and London, in 1911.
5. Frank B. Gilbreth, Primer of Scientific Management (Adamant Media Corporation, 2005). This Elibron Classics Replica Edition is an unabridged facsimile of the edition published in 1914 by D. Van Nostrand Company, New York.
6. Frank B. Gilbreth, *Primer of Scientific Management* (New York: Van Nostrand, 1912), 7.
7. Hiroaki Satake, *Toyota Seisan Houshiki no Seisei, Hatten, Henyou* [The Birth, Development, and Transformation of the Toyota Production System] (Toyo Keizai Shinbunshinhousya, Tokyo, Japan, 1998), 18–19.
8. Shigeo Shingo, *Koujyou Kaizen No Taikeiteki shikou* [Systematic Thinking for Plant Kaizen] (Nikan Kogyo Shinbunsha, Tokyo, Japan, 1979), 12.
9. Interview notes with Katsuya Jibiki, former assistant general manager of Toyota Stamping Department, September 2006, Toyota City, Japan.
10. Flowchart, New World Encyclopedia, http://www.newworldencyclopedia.org/entry/Flowchart.
11. Allan H. Mogensen, *Common Sense Applied to Motion and Time Study* (New York: McGraw-Hill, 1932).
12. Gordon B. Carson (Editor), *Production Handbook* (Ronald Press, 1944, revised 1958), Sect. 11, p. 14.

Chapter 6

Step 3: Generate Original Ideas

6.1 Introduction

The first two steps of Kaizen represent much of the heavy lifting in terms of the analytical work that has to be conducted for developing skills or making improvements. The next part of the Kaizen process is Step 3, generate original ideas, and involves synthesis as much as or more than analysis (Figure 6.1). Subsequently, this phase is quite different in terms of technique as well as approach when compared to the two previous steps. While the act of analysis is generally well defined and even somewhat prescriptive, the opposite is true for generating original ideas. The best ideas sometimes seem to come from nowhere. Occasionally, the act of analysis itself generates automatic ideas for improvement. When insights are not developed during the analysis phase, however, creative thinking needs to occur to generate potential solutions. In this chapter, we share some techniques for stimulating original ideas and synthesizing your solutions.

In this chapter, we cover some of the methods that Toyota successfully used over the years to augment the critical thinking phase of analysis with idea generation. Ideas generally can involve applying different thinking patterns and approaches. Please keep in mind that Kaizen by itself does not supply "answers" for individuals to "cut and paste." Kaizen, much like problem solving or the scientific method, is a structured thinking discipline for developing new and better ways to conduct processes. Individual effort and persistent thinking is required to create new ideas and alternatives. We outline here some of the major concepts employed by Toyota over the years in Kaizen training to help teams get past mental roadblocks and generate original ideas for improvement.

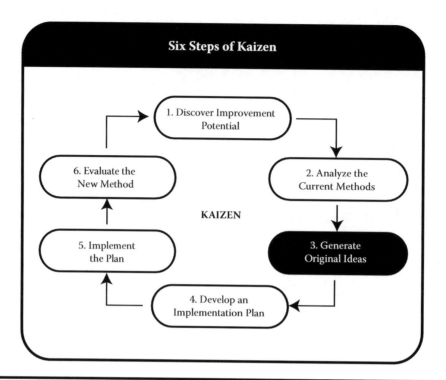

Figure 6.1 Six steps of Kaizen (Step 3).

6.2 Key Concepts Regarding Idea Generation

To help individuals and teams learn to generate original ideas for improvement, we have compiled some of the more common techniques. In reality, there is no secret technique or method that will generate answers for you. Great insights and ideas often spontaneously occur in individuals, and that process cannot always be fully explained. Idea generation sometimes occurs as a result of deep investigation into subject matter. At other times, great ideas seem to appear from nowhere. We discuss some of the common roadblocks to creativity, relate some general rules and observations regarding teams and creativity, and discuss some methods for helping to spur original ideas.

6.2.1 Common Roadblocks to Creativity

Before discussing how to help stimulate and spur creative thinking in individuals and groups, it is worth first reviewing common roadblocks to creativity and improvement. Awareness of these common pitfalls can help you avoid wasting time in many cases and help others when they get stuck in terms of idea generation. Specifically, we outline five typical roadblocks that are common to most environments. There are, of course, other impediments, but these are the most common ones from our experience. We then provide general points of advice for helping to foster creativity in teams and present some specific techniques that you may find useful.

6.2.1.1 *Force of Habit*

One of the first and most common problems in dealing with idea generation is overcoming the force of habit. Humans are fundamentally creatures of habit. We are most comfortable when we are repeating familiar patterns in our lives and daily work routines. *Kaizen* fundamentally means "change for the better." The act of change involves some degree of both courage and creativity to alter the status quo. The definition of *insanity* as Albert Einstein once remarked is to repeat the same process over and over and expect different results. When practicing Kaizen either for individual skills development or for producing results, remember to challenge the status quo. Altering the process in some way is required to create "change" and drive "improvement" (i.e., Kaizen).

6.2.1.2 *Preconceptions*

Preconceptions about the process or end result are also a related form of blockage when driving improvement activities. Often, we have filed away mental notes or statistics that we have heard in the past and assumed to be true. Maybe those bits of information were true in the past regarding a process, the customer, the design, or the supplier. Over time, however, situations change and open up opportunities for improvement. In Kaizen, you must be willing to suspend previous judgments or opinions and test them again from scratch.

6.2.1.3 *Common Sense*

As strange as it may sound, common sense can also be a powerful blocking force in terms of improvement. For example, it was "common sense" that making more parts always leads to greater efficiencies and cost improvement. It is only when you put on the "uncommon lens" of avoiding overproduction that various wastes of overproduction start to become clear to observers.

A similar example in the history of Toyota pertains to setup reduction efforts and changeover work. In manufacturing in the 1950s, it was normal for changeover work on stamping machines to take anywhere from one to four hours depending on the size of the machine. Given this assumption, it is normal to want to avoid changing stamping dies due to the loss of run time incurred on the press. However, the assumption is flawed in this case. Changeover time is not fixed and can be shortened with work. Change time was shortened inside Toyota from the levels discussed to a companywide average of 15 minutes in 1962. By 1973, that average was down to less than 3 minutes per machine.[1] When this short time is possible, it makes great sense to change over stamping dies frequently to meet changes in customer demand and to reduce inventory levels. Challenging assumptions and common sense is often a big part of Kaizen.

6.2.1.4 Not Invented Here Syndrome

Sometimes companies are proud of their history and traditions: "We did not invent that method here, so we don't want to do that type of thing." Some degree of pride is healthy and normal. Excessive pride, however, is arrogance and one of the seven deadly sins in many different cultures. The ancient Greeks, for example, considered "hubris" or excessive pride as a dangerous sin that, when left unchecked, led to the downfall of even the most powerful individuals.

When you stop and review the concepts presented here, not much was truly "invented" by Toyota. In the arena of Kaizen, time study, motion study, and work analysis are all items developed chiefly in the United States and other countries in the early 1900s. Takt time is a concept Toyota borrowed from German aircraft manufacturing. Pull systems replenishment methods have parallels in U.S. supermarkets. The list is long. If Toyota had not been open to other ideas from outside the company, it never would have reached the heights that it has achieved in its respective industry.

6.2.1.5 Emotion

The last powerful blinding force that we mention is that of sentiment or emotion. The human brain uses both logic and emotion to form opinions and then action. Often, emotion is a far more powerful and dominating effect when it comes to challenging current methods or the status quo. Fear is a specific type of emotion that often comes into play. This sort of impediment has to be identified and dealt with at various times in working with individuals and work teams.

Emotion has its role in driving improvements; however, it needs to be channeled and harnessed in a proper way. For example, it is acceptable to be excited and passionate about wanting to change things for the better. Sakichi Toyoda was motivated to develop a better loom to ease the burden on his mother, who operated manual looms. The search for a better way to improve our situation or the situation of a group of employees is a powerful way to harness emotion.

Conversely, we must strive to avoid negative defeatist thinking. Negative emotional thinking can stop improvement work dead in its tracks. Negative thinking can form an invisible web that robs teams of power and stifles creative thinking. The key in Kaizen is to follow the six-step process outlined in this workbook and to apply the outlined thinking patterns. Inside Toyota, managers often talk about the need for developing the "3 Cs" in employees and leaders at all levels. The Cs refer to challenge, creativity, and courage. *Challenge* means being willing to question the status quo and look for better ways. *Creativity* refers to the process of thinking differently and not merely clinging to the ways of the past. *Courage* means the willingness to test your ideas and learn from trial and error. We suggest that you strive to model the 3 Cs when you implement Kaizen.

6.2.2 *General Advice Regarding Creativity and Teams*

Having covered the standard roadblocks that often stifle creativity, let us now shift gears and cover some points of general advice regarding idea generation. The following six practices are important to understand before embarking on idea generation in Kaizen or any other similar activity.

6.2.2.1 *Separate Idea Generation from Judgment*

The first critical concept to keep in mind when promoting creative thinking and original idea generation is to separate idea generation from judgment. The human brain is often quick to pass judgment on ideas and tends to react negatively as a first reaction. It is easier to think of why something will not work than it is to understand how and why it might work. In Kaizen, you need to establish the discipline of avoiding rushing to judgment too soon. Potentially good but uncommon ideas will get trampled by common sense or negativity if you are not careful.

6.2.2.2 *Generate as Many Ideas as Possible*

As a related next step in this process, we also recommend generating lots of ideas first (i.e., quantity) and then worry about practicality (i.e., quality). In reality, of course, quality of the idea is the most important attribute at the end of the day. Bad ideas rarely make for good Kaizen. However, if quality is used as the first-pass filter on thinking, then the human brain tends to become cautious, and individuals proceed with less creativity. The urge to state only "good ideas" makes us conform to existing norms with which we are familiar. The unintended consequence is to shut down the parts of our brain that are trying to think of something new and unusual that might work better.

When implementing Kaizen, make sure the order of these first two steps is followed if you truly want to invite some creative thinking from individuals. If the atmosphere is open and nonjudgmental, then you will tend to get more ideas out on the table. In the end, many ideas will be discarded, but the odds of generating a new and creative idea are better if people are not concerned initially about conforming to existing standards or rushing to judgment.

Here is a sample exercise that you might want to try to review the first two points. Take out a large paper clip and hold it up in clear view for everyone to see. A binder clip or other item would work as well. Tell everyone that they have five minutes to list as many different uses of the paper clip as they can imagine. The person with the most ideas wins the exercise. Many immediate examples, such as a key chain, book marker, hair braid, and the like, will come to mind. Tell the audience that you are looking for over 20 ideas and challenge them to be creative. There is no right or wrong answers for the exercise. This example is just a way to loosen up the mind and the creative thought process.

6.2.2.3 Think from Different Angles

Thinking from different angles is another good way to help generate original ideas. Often, individuals start with preconceived notions and thoughts about a given situation. Sometimes, we need to lose our assumptions and start with fresh perspectives. Here is an example exercise of what we mean. Figure 6.2 shows nine dots in a box. How might you connect all nine dots using only four lines and not lifting your pen at any point in time? Most people are unable to figure out a solution. There is a way to connect all nine dots and not break the rules. The example in Figure 6.3 is how all nine dots can be connected.

As you can see, the answer requires thinking "outside the box" and dropping the preconceived notion that you cannot go outside the cube of the box. Teams and individuals often trap themselves into thinking that they cannot go outside traditional areas of consideration. However, many solutions like this one or others require the courage to think from a different angle.

6.2.2.4 Combine Ideas with Others

There is an old saying that "two heads are better than one." The number of ideas that can be created by a group is far greater than the number that can be created by one individual. The paper clip exercise is a good example. One individual on his or her own may come up with a dozen examples of applications for the paper clip in five minutes. The entire group might come up with two dozen different ideas or more in some cases. The key point to take away from this activity

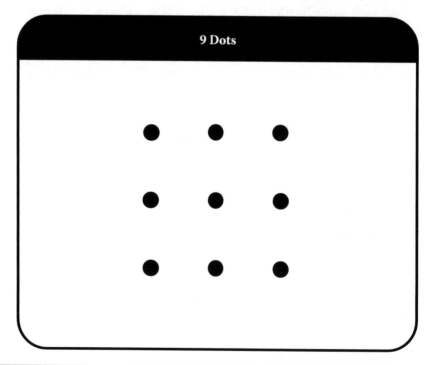

Figure 6.2   Nine dots.

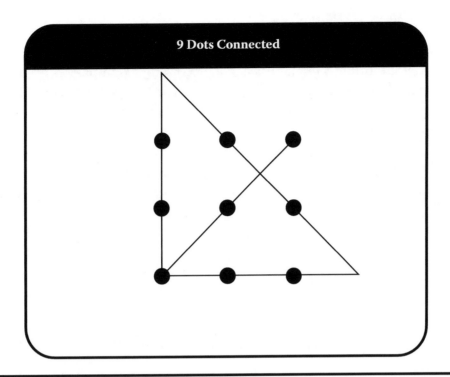

Figure 6.3 Nine dots connected.

is that working in collaboration can often increase both the quantity and the quality of ideas generated. Be sure to involve different appropriate parties when seeking to generate new alternative ways of doing things.

6.2.2.5 Review Previous Analysis

Often, it helps to revisit the analysis you have conducted to generate original ideas. The lenses of time, motion, work elements, flow, and so on can be powerful guides in assisting your thinking. Take time to review any analysis conducted that sheds light on the problem you are solving. Detailed analysis often brings problems into crystal clear focus. Once problems or opportunities are clear, then solutions tend to jump out more easily. For example, just hearing that a job is "difficult" does not do much to generate insight. Seeing in detail that moving a 20-pound finished goods container from Point A to Point B is problematic, however, jump-starts the thinking process. Can a simple gravity feed conveyor be used, or should other potential angles be considered? Clear problem detail makes for clearer idea generation. Fuzzy detail yields low-quality insights.

6.2.2.6 Synthesis Ideas

The methodology outlined in this workbook is heavily stilted toward analysis or reducing things into smaller units for study and conjecture. The Greek root of the word *analysis* is in fact *analyein*, which roughly translates as to "break things up." This technique is an excellent method for studying how things work and developing new insights.

It is equally important to remember that, according to Immanuel Kant in his work *Critique of Pure Reason*, knowledge is created by two functions: analysis and synthesis. Analysis tends to get all the attention in studying current methods, but synthesis is equally important for the creation of ideas and hence Kaizen. The Greek root for synthesis is *syntithenai*, which is Greek for "to put together." Creative idea generation often stems from putting two different pieces of information together and generating a new and original idea. For example, obtaining bolts in any assembly operation is a monotonous task. Walking two steps and obtaining exactly four each time is neither value added nor easy. Having a simple device that counts four bolts for you and positions them near the point of use, eliminating the need for walking, is an interesting solution if possible. Brainstorming a fastening method without bolts is even better. Regardless of the example, it is important in any Kaizen not only to break things down for detailed study but also to put things back together in creative different ways that can lead to improvement.

In summary, practice adhering to these six principles for idea generation and you should be off to a good start. In reality, no one thing sparks or drives creativity in people. What works for one person often may not work for another. Extroverted individuals may gain more creative spark from debating ideas with others in a group setting. Introverted personality types may develop creative insights from making drawings or conducting deeper observations on their own. Random thought association techniques work well with other groups. Experiment with different techniques yourself, and in group settings to see what works best for your case and plot a course for idea generation.

6.2.3 Methods for Developing Ideas

From our experience, most truly creative ideas spring from the mind and not a check sheet of ideas. However, as you might guess, various check sheets and lists have been created over the years to aid participants in thinking about creative new ways to do things. We suggest the following lists as mental checklists to ascertain if you have covered all the areas discussed next. Do not expect answers to leap magically off these pages as solutions to your respective situation. Several such checklists are provided for you to consider.

6.2.3.1 Osborn's Checklist

Alex Osborn is regarded as the father of classical brainstorming. In one of his published works, Osborn created a mental checklist of items for participants to consider when creating new ideas.[2] The checklist outlines seven categories for consideration, with associated questions for each category (Figures 6.4 and 6.5). While it is unlikely that application of this checklist will generate answers for

Osborn's Checklist 1/2

1. Is there any way of reusing what you don't need any more? (Reuse)
   Can rejected and useless items be used for something else?
   Can we think of new way to use goods and materials?
   Can somebody's personality and ability be used somewhere else?

2. Can a similar item be used for something else? (Borrowing)
   Are there any other things that look alike?
   Are there any imitations?

3. Can we change anything? (Change)
   Can we change the color?
   Can we change the shape?
   Can we change the sound?

4. Can we enlarge it? (Enlargement)
   Can we make it bigger?
   Can we make it longer?
   Can we make it stronger?
   Can we make it thicker?

5. Can we reduce it? (Reduction)
   Can we make it smaller?
   Can we separate it?
   Can we compress it?
   Can we make it lighter?

Figure 6.4   Osborn's checklist, page 1 of 2.

your situation, it is a good tool to review and ponder at times. It may drive teams to open their minds and help spur insights or steer thoughts toward neglected areas.

6.2.3.2 Rules for Motion Economy

Industrial engineers also created checklists for examining work and looking for improvement potential. A common technique was the application of Frank Gilbreth's rules for motion economy, which were developed after many years of study. The rules are broken into three respective areas: the human body (Figure 6.6), arrangement of the work site (Figure 6.7), and the design of tools and equipment (Figure 6.8). For review, these figures provide the main points of Gilbreth's different checklists regarding principles for motion economy.

As in the previous instance, we doubt that you will find automatic answers to your problems in these bits of information. However, review the contents carefully, and you may pick up areas for further investigation or analysis. Since these "rules" are from the early part of the 20th century, there is a heavy emphasis on

Osborn's Checklist - 2/2

6. Can we substitute something? (Substitution)
   Can another person be substituted?
   Can other things and materials be used instead?
   Can the work process, power, and place be substituted?

7. Can we rearrange it? (Replacement)
   Can we rearrange the elements?
   Can we rearrange it into some other form?
   Can we rearrange to another layout?
   Can we rearrange the order?

8. Can we reverse it? (Reverse)
   Can we change plus to minus?
   Can we reverse the roles?
   Can we reverse it up/down? Left to right?

9. Can we combine anything? (Synthesize)
   Can we mix?
   Can we combine each unit?
   Can we combine ideas?
   Can we combine elements?

Figure 6.5   Osborn's checklist, page 2 of 2.

manual operations and simple tools. With some extrapolation effort, you can make them apply to almost any situation or create your own list.

6.2.3.3 Further Suggestions for Manual Work

In addition to Gilbreth's rules of motion economy, Toyota developed some more specific topics to consider in Kaizen activities. The Toyota examples mainly apply to Toyota's type of operation and facility. However, you might find some areas for consideration or parallels for your situation. The list may also give you a framework for generating your own internal checklist of ideas to consider. The checklist we have outlined based on Toyota experiences covers multiple basic categories. Rather than outline the entire list here, we outline the categories and place the list in the appendix section of this workbook for interested parties to peruse. (Refer to Appendix 1 for details on these points of consideration.)

6.2.3.4 Review 5W 1H and ECRS

As mentioned in this section, one of the best ways to generate ideas if you are stuck is to go back to the original situation and revisit the analysis of the

Motion Economy - Use of the Human Body

1. The two hands should begin and complete their movements at the same time.

2. The two hands should not be idle at the same time except during the rest periods.

3. Motions of the arms should be symmetrical and in the opposite directions and should be made simultaneously.

4. Hand and body motions should be made at the lowest classification at which it is possible to do the work satisfactorily.

5. Momentum should be employed to help the worker, but should be reduced to a minimum whenever it has to be overcome by muscular effort.

6. Continuous curvilinear movements are to be preferred to straight line motions involving sudden and sharp changes in direction.

7. "Ballistic" (i.e., free swinging) movements are faster, easier, and more accurate than restricted or controlled movements.

8. Rhythm is essential to the smooth and automatic performance of a repetitive operation. The work should be arranged to permit easy and natural rhythm whenever possible.

9. Work should be arranged so that eye movements are confined to a comfortable area, without the need for frequent changes of focus.

Figure 6.6   Motion economy: use of the human body.

problem. Sometimes after thinking, we realize that there might be a better way to analyze the current situation. In other cases, reviewing the detailed analysis with some hindsight helps clarify the problem or opportunity. Once the situation is clarified, answers and alternatives tend to be easier to generate. Reviewing the 5W 1H (what, why, where, when, who, and how) and ECRS (eliminate, combine, rearrange, simplify) framework (Figure 6.9) can be a good way to focus on problems again and generate creative ideas for improvement. The structured thinking process of ECRS is always a good place to revisit during Kaizen.

6.2.3.5 Brainstorming

The last concept that we cover is brainstorming. This technique has achieved such popularity that you might wonder why we do not include it first on the list of ways for creating original ideas. Since it is probably the most widely known method, we chose to leave it for last to highlight some of the older roots and methods for creating new ideas. The process of classical brainstorming, as noted, is generally accredited to Osborn and is now over 50 years old. Entire books exist on this topic, and complete coverage of the method is not possible. For simplicity, we distill the method to the key categories that Toyota practiced and taught in

Motion Economy - Arrangement of the Workplace

1. Definite and fixed stations should be provided for all tools and materials to permit habit formation.

2. Tools and materials should be prepositioned to reduce searching.

3. Gravity feed bins and containers should be used to deliver the materials as close to the point of use as possible.

4. Tools, materials, and controls should be located within the maximum working area and as near to the workplace as possible.

5. Materials and tools should be arranged to permit the best sequence of motions.

6. "Drop deliveries" or ejectors should be used wherever possible so that the operator does not have to use his hands to dispose of the finished work.

7. Provision should be made for adequate lighting, and a chair of the type and height to permit good posture should be provided. The height of the workplace and seat should be arranged to allow alternate standing and sitting.

8. The color of the workplace should contrast with that of the work and thus reduce eye fatigue.

Figure 6.7 Motion economy: arrangement of the workplace.

Motion Economy - Design of Tools and Equipment

1. The hands should be relieved of all work of "holding" the work piece where this can be done by a jig, fixture, or foot operated device.

2. Two or more tools should be combined wherever possible.

3. Where each finger performs some specific movement, as in typewriting, the load should be distributed in accordance with the inherent capacities of the fingers.

4. Handles such as those in cranks and large screwdrivers should be so designed that as much of the surface of the hand as possible can come into contact with the handle. This is especially necessary when considerable force has to be used on the handle.

5. Levers, crossbars, and handwheels should be so placed that that the operator can use them with the least change in body position and the greatest "mechanical advantage."

Figure 6.8 Motion economy: design of tools and equipment.

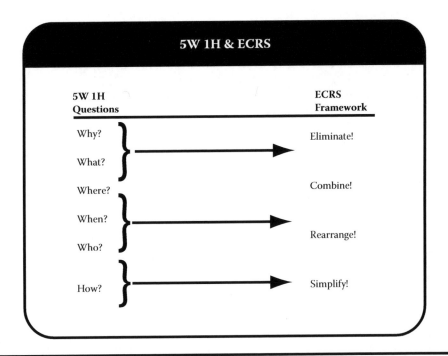

Figure 6.9   5W1H and ECRS.

operations as part of the Kaizen course (Figure 6.10). You may want to revisit the concept in its entirety and expand or contract as your situation may require.

Brainstorming is a process by which we seek to generate new and original ideas for better results. A brainstorming session is a group session in which participants work together to generate original ideas and identify new solutions. Brainstorming seeks to take advantage of the energy of the group to build a chain reaction for idea generation. For simplicity, we established four main rules for brainstorming (Figure 6.11).

The rules are fairly self-explanatory and build on ideas already established in this chapter. For example, free thinking is urged in the meeting, criticism is suspended for the time being, quantity is the initial goal, and working together to synthesize ideas is encouraged.

As a sample process, we suggest something along the lines presented in the discussion that follows, but other ways will work as this is merely a suggested practice to following for those parties looking for a quick-and-easy way to get started. In the following paragraphs, we cover some suggested items for preparation, review a couple of critical success factors, and discuss some sample roles to employ.

In terms of preparation for brainstorming, be sure to select a quiet, comfortable location where participants will not be disturbed or distracted. During the meeting seek to avoid all forms of interruptions such as phones, pagers, computers, and the like. Generating original ideas requires concentration and not scattered attention. The participants should depend on your specific situation. Some mixture of process experts and some degree of people from external areas are generally suggested. Be sure to cover the rules and best practices for

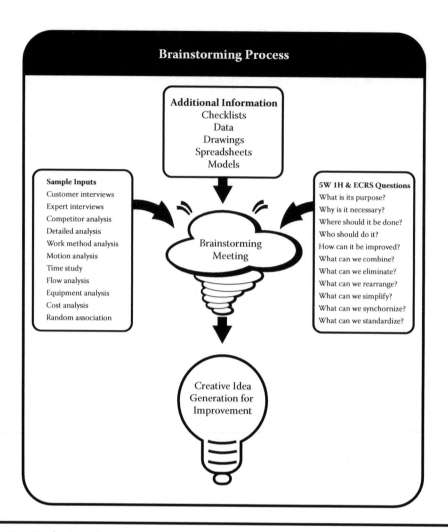

Figure 6.10   Brainstorming process.

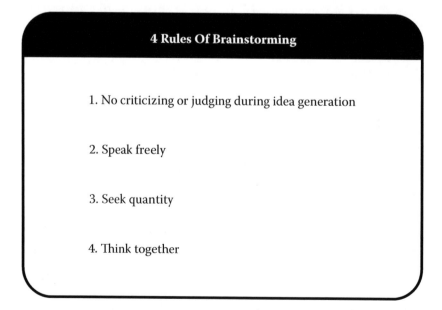

Figure 6.11   Rules for brainstorming.

brainstorming in advance with the group. Limit the time to the practical amount necessary for the group. All-day meetings are not necessarily going to generate more or better ideas than a well-run two-hour meeting.

In terms of generating success, we suggest keeping in mind the following items: Make sure you have a clear theme for consideration. In other words, avoid abstract themes or vague concepts. The mind will focus better when the topic is clear. When possible, communicate the theme of the meeting in advance by a day or two to the participants. There is no harm in thinking before coming to the meeting, and this is generally more productive than waiting to communicate the topic.

Stay away from making judgments or choices part of the theme. It is all right to ask how we might make a product lighter in weight or a process shorter in terms of cycle time or the like. However, do not ask, for example, "Should we use material A or B?" or "What type of new machine should we purchase?" These examples are not real questions and limit the scope to preselected answers.

In terms of running the meeting, we also suggest employing someone as a dedicated facilitator and someone else as a scribe (discussed further in this section). These two roles can be combined in smaller groups but this becomes difficult when a group of seven or more people is present. When there are a lot of people interested in participating or who need to be involved for whatever reason, you might instead want to run two groups in parallel instead of one large group. Comparing and synthesizing the results of the two groups afterward may lead to further creative insights. More participants do not always generate better results. Keep the number of people in the meeting to a number that you can manage.

In terms of facilitation, seek to create an open, cheerful atmosphere. Have participants raise a hand and speak in turns if communication becomes difficult to manage. Make sure that everyone is participating in some fashion. Remember, however, that different personality types might contribute in different ways. Ensure that the four basic rules of brainstorming are followed at all times. Discourage any negative critical thinking that might occur during the idea generation phase. Ideas will be evaluated later in the process. Encourage people to build off of others' ideas or "piggyback" as needed. The facilitator can always pose ideas for consideration when no ideas are forthcoming. This, of course, takes some up-front preparation time on the part of the facilitator.

If the group is large enough (i.e., seven or more people), facilitation can become difficult as multiple ideas and conversations start to be generated. In this case, consider appointing someone to function as a scribe in charge of writing down the ideas. The scribe mainly focuses on the task of writing down the different ideas. This is a big help in facilitation as the facilitator is able to remain facing the group and does not need to turn to write on a flip chart or whiteboard. The ideal scribe is someone who can write neatly and quickly. It helps to summarize and restate things verbally before writing them down to avoid any miscommunication. For large groups, use of two scribes is also an option to consider.

There are many other techniques to employ in brainstorming or points of advice for when teams are stuck and need assistance moving forward. We suggest finding a facilitator's handbook for brainstorming in these sorts of advanced cases that you might need to handle. The majority of the time in Toyota's style of Kaizen and idea generation, the techniques outlined here were sufficient to spur creativity and drive improvement.

The last aspect of brainstorming and idea generation that we comment on is that of idea evaluation. Eventually, critical thinking is required in the process and must be dutifully employed. Bad ideas, no matter how popular they might be, do not make for good Kaizen. After idea generation is complete, there are some general rules to employ and keep in mind.

If a large number of ideas is generated, they need to be pared down in some manner. Nominal group technique or voting can be used to pare down and eliminate some of the options that have been created. The danger in such a method is that a good idea understood by only a small number of individuals can easily go to waste.

An alternative method is to rank the ideas generated by categories by different scoring mechanisms. For example, seven different, mutually exclusive ideas may exist for improving a process. List the seven solution sets and score them against the following categories: cost, difficulty, impact, effect on quality, productivity, customer requirements, and so on. Scoring can be done in terms of high, medium, and low or on a basis of 1 to 5. Adjust the scoring criteria and scoring mechanism to fit your needs.

In closing, however, remember that the highest-scoring or most popular idea may not be the best answer in terms of actual improvement. Often, ideas win because they are popular or a few dominant personality types in the room champion these ideas. In Kaizen, make sure that all ideas and solution spaces are adequately explored before choosing a new way of doing things. In practice, it is always best to organize a trial in which competing ideas can be tested against one another head to head for comparison. Time, cost, and difficulty often stop us from conducting such rigorous evaluation testing. However, when time allows, we suggest that you pilot the competing ideas as is feasible under the circumstances. Make the final decision as objective and quantitative as possible and not just a subjective opinion or preference.

6.3 Summary

This chapter represents an important step forward in idea generation and decision making for Kaizen. Examining current methods in Step 2 is highly analytic. This next step involves creative thinking to generate new and better solutions. By definition, this process involves synthesizing information as much as or more than it involves analyzing. Taking adequate time during this idea generation and decision evaluation phase is of critical importance.

6.4 Homework Assignment

There are multiple techniques and topics to master in this section of the workbook. No single technique or checklist will ever do your thinking for you. The way to true and original Kaizen is by focusing your time and attention on new and better ideas. The mantra used in Toyota for decades was "creativity before capital" in spurring teams to think about new solutions. We suggest trying all of the techniques in this chapter as time and conditions permit. View all of these points of assistance as aids for thinking and do not fall into the lazy trap of thinking that there are simple solutions somewhere in some book. In Toyota, it was often commented by various leaders that the T in TPS (Toyota Production System) really stood for "thinking" and not just Toyota. Keep the principles from this chapter in mind during Kaizen, and we predict that you will have an easier time generating creative solutions.

Notes

1. Hiroaki Satoshi, *Toyota Seisan Houshiki no Seisei, Hatten, Henyou* [Birth, Development and Transformation of the Toyota Production System] (Tokyo: Toyo Keizai Shinbunshya, 1998), 18–19.
2. Alex F. Osborn, *Applied Imagination: Principles and Procedures of Creative Problem-Solving* (New York : Scribner, 1957).

Chapter 7

Step 4: Develop an Implementation Plan

7.1 Introduction

Once the analytical work and idea generation phases are complete, it is time to make a Kaizen plan. Kaizen plans come in all shapes and sizes. Often, the best plan is "no plan" at all, just getting things done right on the spot. In other instances, that approach is not feasible due to the difficulty or lead time involved in getting changes approved or materials prepared and so on. In this brief chapter, we discuss some key points and a couple of different ways of thinking about your Kaizen plans and how to keep the ball rolling. As Benjamin Franklin once remarked, "By failure to prepare, you are preparing to fail." This chapter will help make sure that you are planning for success and obtaining the best possible results in your Kaizen plans (Figure 7.1).

7.2 Key Concepts

In this section, we first look at six important points to address in moving forward with Kaizen and then move on to three general cases. The six points are not a complete list of all things that you will need to consider but rather an introductory grouping of general things that all parties will probably face. Specific instances in your area will have to be layered on top of these as needed. We provide some simple, yet practical, thoughts on writing short, effective Kaizen plans in addition to reviewing the key concepts.

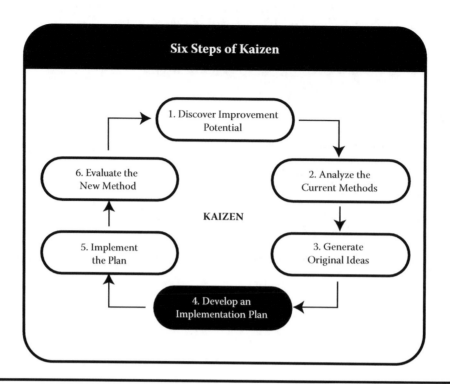

Figure 7.1   Six steps of Kaizen (Step 4 focus).

7.2.1 Six Points for Consideration in Planning

The first key concept involves reconciling your end goal with your means of achieving that goal. Kaizen is about "change for the better" and not merely change for the sake of change. It is critical to establish that the ideas proposed in Kaizen outweigh the costs or difficulty of attaining that goal. For example, if we target a productivity improvement of 15%, which represents $1 million in savings, yet the plan to get there costs the company $2 million, then this is not Kaizen. It is the opposite or "change for the worse." Ideally, in Kaizen the costs associated with method-based changes are so trivial that substantial approval is not required. However, in some cases careful cost-versus-benefit analysis must be conducted. Follow existing rules and guidelines in your company for obtaining approval and justifying improvement suggestions.

A second related key point is that Kaizen should initially focus on method-based changes to the way things are currently done. This is important for several reasons. First, it is easier to change methods than it is to change product or equipment design. Often, the latter involves extensive approval and validation testing, which can take quite a while. Cost is another reason to favor method-based changes first as an initial starting point. The mantra of "creativity over capital" should be emphasized to every area hosting Kaizen activities. Changes in method also lead to better insights about how future products or machines might be designed. For multiple reasons, we advise erring on the side of method-based changes before capital-based ones.

A third key point is to ensure that improvements projected from the Kaizen activity are truly system-level Kaizen and not merely local improvements that will have minimal effect. For example, imagine a series of process steps organized in a linear flow. The final process is the limiting or bottleneck process with respect to meeting customer demand. Implementing Kaizen on processes preceding this unit may look nice on paper but will not produce any more products for the customer at the end of the day. We have observed that this concept is often overlooked in many cases, and it should be carefully reviewed. Focus Kaizen activities on key points in the system that will unlock greater system performance. Do not just cherry-pick easy processes to fix and then make grand claims of system-wide efficiency improvement.

All Kaizen plans should include proper analysis to ensure that both safety and quality are not harmed by any proposed changes. As part of your Kaizen planning process, you need to make sure that safety and quality remain at the same levels as the previous method and ideally are both greatly improved in the new proposed method of doing things. The best practice we can suggest is to pilot changes in advance of implementation. As a fallback position when this is not possible, verify the changes with all affected parties in advance to help ensure no ill effects will come from the implementation of the new methods. We cannot stress enough the critical importance of piloting the changes and measuring the difference, especially in regard to safety and quality.

Often, however, piloting changes is difficult and not feasible. Or, sometimes only partial testing can be accomplished. Because of these realities, in all cases it is wise to put in place adequate backup plans for dealing with worst-case scenarios. Modifications to work or machines might yield initially promising results with regard to quality or safety, for example, but later prove to be unsustainable or worse than before regarding other critical dimensions. In these instances, the proper measure is to temporarily suspend the Kaizen action items and revert things back to the previous state. This does not mean giving up; it means revisiting your analysis, original ideas, and the effects that were generated. Trial and error are part of Kaizen, and it often takes multiple iterations to get something just right. Giving up at the first sign of trouble is unacceptable behavior. A high degree of perseverance is required. Take adequate precaution in Kaizen planning to ensure that things can be returned to normal fairly quickly in the unlikely case of poor results. Failure to do so not only is the opposite of Kaizen but also is fodder for critics of any future Kaizen activities.

The last key point we mention is to consider the effect of any changes on the group, certain individuals, and the company. Input and involvement from team members can often be a great way to foster a sense of belonging and contribution. Participation in Kaizen activities is an excellent way to develop employee potential and foster improvement. Take time to ensure that ideas from employees inside work teams are included as well as ideas from other constituents. Remember, what may seem small to you can be large in the minds of other people.

Keep in mind, however, that conflicting signals might be received from different departments when cross-functional work is substantially changed. A more efficient layout and delivery of materials to a department of 20 people might yield great improvements to the production group. However, the material-handling group might see this as more work for their constituents. In the end, disagreements like this have to be resolved with what is the best for the company and the most efficient total system solution. Take time to deal with these different parties as part of Kaizen and do not let such cases become negative points of contention.

7.2.2 Implementation Cases and Planning

As we remarked, the best case in Kaizen is not to require extensive planning on paper and instead rely on analysis and quick decisions to keep things moving. This mode of quick Kaizen, which we often refer to as "point Kaizen" or fairly simple Kaizen, exists entirely within your realm of control. In other words, this might be called the "just-do-it" school of Kaizen.

Examples of this type of activity might be conducting the 5 Whys to ascertain why a machine is failing or why a quality problem is occurring. Study of motion or time or work elements might highlight simple local gains as well that make sense to capture without delay. We wholeheartedly recommend implementing these types of improvements as quickly as possible. Change for the better is contagious in a positive way. One improvement often opens the way to another. Change for the better opens eyes to what is possible rather than what is not possible. Change for the better instills confidence and vigor in an organization. Failure to make simple changes for the better is the same as procrastination on the individual level. It not only accomplishes nothing but also leads to a negative sense of self-worth. There is an anonymous quote about the genius[*] of boldness that applies to Kaizen.

> Until one is committed, there is hesitancy, the chance to draw back—Concerning all acts of initiative (and creation), there is one elementary truth that ignorance of which kills countless ideas and splendid plans: that the moment one definitely commits oneself, then Providence moves too. All sorts of things occur to help one that would never otherwise have occurred. A whole stream of events issues from the decision, raising in one's favor all manner of unforeseen incidents and meetings and material assistance, which no man could have dreamed would have come his way. Whatever you can do, or dream you can do, begin it. Boldness has genius, power, and magic in it. Begin it now.

---

[*] Sometime attributed to Johann Wolfgang von Goeth, but of uncertain origin.

Some types of Kaizen might not be possible to complete on the spot or in a single day. Sometimes, we refer to these as *flow Kaizen* or *system Kaizen*. These are activities that cross boundaries and hence may take longer to obtain approval or validation or be implemented. When it is possible to get things done quickly, move in that fashion. When that is not possible, then some degree of planning and organization is required to coordinate activities.

There is no rule that Kaizen must occur over a set period of time. In Toyota, for example, small Kaizen occurs every day in the course of natural work. Medium or larger types of Kaizen activities often take days or weeks to accomplish. Less-frequent cases involving design changes, equipment modifications, and so on might take weeks or even months to complete. Outside of Toyota, a common pattern has been to hold five-day event-type workshops for the purpose of implementing Kaizen. As we pointed out in the background section of this workbook, such an implementation pattern is not that common inside Toyota Motor Corporation. The point we would like to make is that the duration of your Kaizen event is not the important point of focus. The quality of the result and the frequency with which you are able to implement Kaizen are more important to your overall success.

In this vein, it is important to become skilled in simple yet effective communication patterns so that progress will continue with minimal delay during extended Kaizen activities. Short meetings focusing on what needs to be done, by which party, by which date, and to what extent need to occur at appropriate intervals. If these types of brief communication meetings do not occur, then progress in implementation often lags. Simple graphs and charts can make this communication occur more efficiently. Figure 7.2 indicates one such example.

Work Plan Update Example

Kaizen Focus Point	Action Item	Responsible	Due	Status
1. Dirt and contamination	Daily 5S & PM tasks	Tony (T/L)	11/2	Conducting daily.
2. Set up reduction	Tool storage cart	Tony (T/L)	11/4	In place and evaluating.
3. Walk distance	Move stations 3 & 4	Tony (T/L)	11/4	Planned for this weekend.
4. Abnormal noise in machine	Spindle bearing check	Ed (Maint)	11/5	Loose bearing cap. Tightened.
5. Parts hanging on exit conveyor	Adjusted angle	Ed (Maint)	11/5	Complete. No hangs reported.
6. Wait time in mid-cycle	Changing PLC program	Mary (Eng)	11/5	Complete. 3 seconds shorter.
7. Poor part presentation	Rearranged layout	Janet (T/M)	11/9	Complete. 4 seconds saved.
8. Poor lighting in work area	Replace lamps	Janet (T/M)	11/9	Complete. Much better.

Figure 7.2   Work plan update example.

7.3 Summary

In this chapter, we have covered the bare essentials of making a Kaizen plan. The goal of Kaizen, however, is improvement and not writing plans. When possible, make changes quickly and effectively as the situation allows. When Kaizen activities extend over several days or weeks, be sure to create effective plans for communication and tracking purposes. The process of writing a plan is good practice and can be useful again later in communicating the story of the Kaizen activity as well as the results.

7.4 Homework Exercise

Practice creating a Kaizen plan for either recent work you have completed or work you are in the process of completing. Overall, the plan should cover the key concepts we have discussed in this chapter as necessary to ensure success. In terms of structure and communication, the plan might take either of the forms we have listed. Your plans may be brief tables of information such as displayed in the Workplace Update Example on the previous page. Alternatively for larger projects, you may wish to write a full-sized 11 x 17-inch A3-sized report. Typical components of an A3 report include 1) the background of the project, 2) the current state of the process, 3) your goal of Kaizen, 4) analysis of the current state, 5) action items for implementation, 6) a way to check your results, and 7) any follow-up items.

Chapter 8

Step 5: Implement the Plan

8.1 Introduction

In this chapter, we will discuss some aspects of Kaizen related to implementation plans. Each company and implementation situation is unique, so there is little that can be shared in terms of details that is universal in nature. However, there are some key points of consideration we would like to communicate and suggest that you focus on during implementation. Failure to attend to these details can often cause Kaizen activities to have a less-than-desired impact or even to run into unexpected barriers.

8.2 Key Concepts

In this section, we cover three particular aspects relating to Step 5: Implement the Plan (Figure 8.1). The first key point covers the importance of thorough communication with affected parties in your organization. The second point addresses the importance of instruction and follow-up action items. The final point is related to creation of a positive atmosphere and attitude toward implementing Kaizen.

No matter how good a Kaizen plan is, the results are only as good as the quality of the analysis on which it is built and the quality of the implementation. Having a good strategy is of no use unless it is executed properly. Plans on paper have to translate into specific and effective action items on the shop floor that result in change for the better. One area that often derails Kaizen implementation is failure to communicate adequately with all parties affected by the change. All implementation plans have to be communicated and sold to some extent.

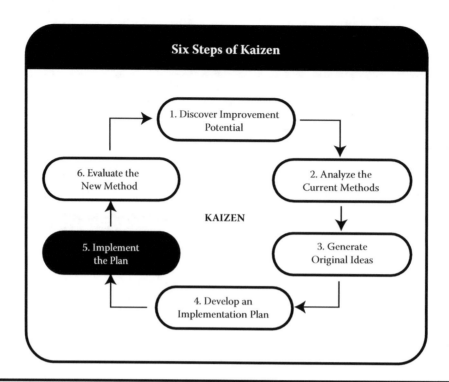

Figure 8.1   Six steps of Kaizen (Step 5 focus).

Let us consider a fairly simple change to a production sequence. The production layout and work flow are altered; the nature of some work elements and machine cycle times are shortened. Other nonessential activities are eliminated, and non-used items are moved into storage. The anticipated gains are a 12% productivity improvement, a 25% reduction in scrap, and a 33% reduction in production lead time. Safety and morale are expected to improve, but this area had no particular problems in that dimension at the beginning for the sake of this discussion.

What tends to hinder implementation is ineffective communication with affected parties. Changes in operations almost always have an impact on quality, maintenance, materials, human resource groups, and other vital functions. Of course, some of this can be deflected by making affected parties part of the change process. Still, not everyone from every department affected by change can be involved. To mitigate misunderstandings, it is necessary to conduct what often feels like overcommunication of the planned changes. In addition to the obvious implementation steps that must be accomplished, it is necessary to communicate the who, what, where, when, why, and how of change plans. The bigger the change, the more communication that generally needs to occur.

For the purpose of communication, simple tools such as A3 reports (refer to example in Chapter 7 on making a Kaizen plan) can be extremely useful. Short (10 minutes or less) updates can be given to various groups as needed. When face-to-face communication is not possible, then distribution of an A3 Kaizen plan can often function as an effective communication medium.

For the most part, individuals are fairly rational beings. Logic and emotion both guide our thinking, but logic tends to win out in the long run for most people. Part of being a good Kaizen leader is skill in leadership and communication. Communicate upcoming implementation and request assistance from groups that are affected by any change. Extend to other parties the same courtesy that you would expect them to extend to you, and implementation will occur much more smoothly. For example, if you are moving equipment as part of an activity, don't just ask maintenance, for example, to relocate the machines and various utility hook-ups. Explain overall what is going on and why it is important. Take time to find out if the new layout proposes any problems such as hindering maintenance access space to machines. Find out if maintenance has some ideas it would like to incorporate as well.

A second critical item that holds back implementation and attainment of results pertains to effective training and instruction. All too often, change is produced by a few key members who have been afforded some time to work on improving a process. That small team or group is able to make the improvement work, but the net result is less than expected when transferred to the entire department or across shifts.

One reason this occurs is failure to have a job instruction training plan in place as part of implementation. When you change a process, this often triggers the need for training to occur for affected parties or members of the team on opposite shifts and so on. Failure to conduct this training can cause frustration and resentment and limit the results of the Kaizen activity.

In general, we can make the following statements about Kaizen and training. First, always have a training plan in place for proper job instruction when a process is changed. The plan needs to update any documentation, such as job breakdown sheets, work instructions, or standardized work used in training. In addition, the training needs to address all affected parties across shifts and departments that are affected.

Second, time often needs to be allowed for changes to sink in and take full effect. Often, there is a slight decline after some initial improvement, or anticipated changes do not appear as expected. Part of the reason may be that it simply takes time to adjust to new routines. Changing familiar habits and routines can be counterproductive in the short run. This does not always mean that the Kaizen content is poorly conceived. Rather, it may take a couple of weeks for muscles to learn new jobs and minds to become familiar with new patterns. If the Kaizen content is solid, then make sure to give it adequate time to surface results. Investigate which areas are giving people trouble and remove those obstacles. Often, unintended consequences crop up and need further investigation and attention. Do not naively assume that every change will always immediately generate positive results. Sometimes, there is a lagged effect, and the reasons for the delay need to be understood and handled.

The last comment that we make regarding implementation is that it should occur in as positive and energetic an attitude as possible. Leadership and attitude play a large role in helping to ensure positive results in implementation. In team sports, it is the responsibility of the coach to ensure that team members play to their fullest ability and display good teamwork. Similarly, in Kaizen it is the manager or change leader who is responsible for helping to make sure that everyone is aware of the upcoming change and is clear why it is occurring. A positive and honest attitude is always an effective ingredient in the change process. Conversely, a negative and disingenuous attitude will stifle most teams and progress.

8.3 Summary

Implementation is essentially the "do" phase of the plan-do-check-act management wheel. Often, companies are good at jumping into implementation of various ideas. As we demonstrate in this workbook, it is vital to effectively analyze operations for improvement and generate original ideas for improvement. Making a Kaizen plan and implementing the plan are vital parts of the process. Adhere to the advice in this section and you will avoid some of the more common pitfalls we tend to observe.

8.4 Homework Exercise

During implementation of Kaizen-style activities, take care to follow the advice outlined in this section. In advance of change, communicate effectively with all affected parties and make sure vital feedback is obtained in advance of and during implementation. Always make a training plan as part of implementation. Job instruction training is a great place to look for simple answers in this arena.

Chapter 9

Step 6: Evaluate the New Method

9.1 Introduction

The final stage of the six-step Kaizen process is to evaluate the results of the action items performed in order to verify the actual level of improvement. Without measured improvement, there is no Kaizen. Change for the sake of change is not improvement and is a tremendous waste of resources. In this discussion of the final step of Kaizen, we outline the simple, yet effective, ways to verify the results of Kaizen activities. We also cover the importance of standardizing work practices and following up to ensure that gains are solidified.

9.2 Key Concepts

There are several important aspects of Kaizen that cannot be compromised. Arguably, different forms of analysis can be utilized, and different methods for generating creative improvement ideas can be conducted as well. Implementation plans can take various paths to completion. The final aspect of verifying results must be conducted or the entire process of Kaizen is at risk (Figure 9.1).

One of the most common failures we observe in Kaizen activities is the failure to establish that tangible gains have actually been made. In problem solving, this step is typically called checking your results. In the scientific method, this concept is often referred to as verifying and independently repeating results. In other words, the goal is to establish beyond any doubt that linkages between

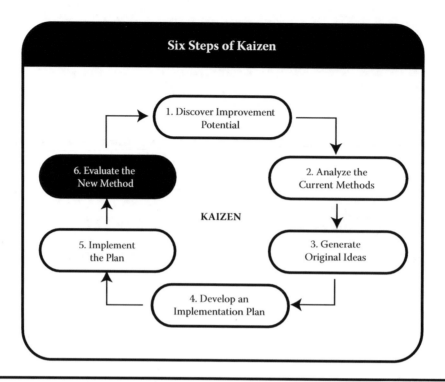

Figure 9.1 Six steps of Kaizen (Step 6 focus).

cause and effect are well understood, and measureable results are obtained from certain actions with a high degree of confidence.

Unfortunately, in Kaizen it is easy to get caught up in the excitement of change and be satisfied just with the amount of activity that has occurred. In other words, activity is mistaken for accomplishment. For example, if an area is visibly cleaner and machines, material, and people are in different locations, it may honestly appear to be a superior arrangement of the work site. But, is it in reality measurably better? This is the final and ultimate test of Kaizen. How do you prove that conditions before and after are measurably "better" and not just "different" to the casual observer? This is one of the hardest parts of the Kaizen process to instill in participants and to entrench in any organization.

Twenty years ago, Russ Scaffede was one of the first Americans to attain the level of vice president in the Powertrain Operations for Toyota North America. During his tenure, Scaffede had the remarkable opportunity to be mentored by Fujio Cho, who at the time was head of North American operations for Toyota. Cho is one of the few remaining disciples of Taiichi Ohno and eventually became president and chief executive officer of Toyota worldwide. As the vice president of operations, Scaffede often led tours of the engine plant that he managed. On a monthly basis, Cho insisted on viewing improvement activities that were under way. Team members would conduct short presentations on the shop floor using flip charts and discuss their improvements.

Cho would comment on the Kaizen activities to the team members and after the tour provide advice and coaching points to managers such as Scaffede. One of the first lessons Cho had to impress on managers such as Scaffede was the importance of "standards" and the necessity to measure before and after conditions to demonstrate improvement.

The word *standard* has several definitions in most dictionaries. One of the most fundamental definitions refers to standards as a "basis for comparison." In Kaizen, we are always looking for this basis for comparison. Safety, quality, cost, productivity, delivery, and other areas are all potential avenues for measuring Kaizen activities. The critical and often overlooked feature, however, is having an absolute standard that can be used to compare before and after situations and make a quantitative statement regarding whether improvement has occurred.

Unfortunately, this practice does not occur naturally in most individuals. The normal tendency is simply to show work in process or action items completed and to consider that is all that can be expected. Kaizen or any improvement activity, however, must show progress toward a goal. Furthermore, it must objectively state whether improvement has truly occurred. This process is of paramount importance; however, it need not be complicated. In fact, we suggest keeping it as simple as possible to establish reasonable proof that improvement has occurred. Figure 9.2 is one such example involving quality of product through a process. In the process, a new type of bearing was employed, and a new type of coolant was added as well. With knowledge that no other changes had occurred to the process, it is reasonable to conclude in this case that rejects were reduced and improvement had occurred.

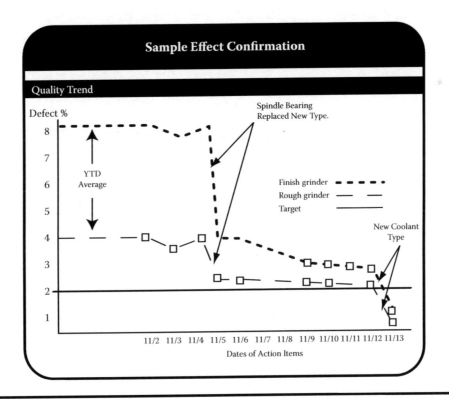

Figure 9.2   Sample effect confirmation.

Sample Before and After Metrics

	Before Kaizen		After Kaizen
Productivity	5PPH*		8PPH*
Quality	2% Scrap		0.5% Scrap
Inventory	7,000 Pieces		17 Pieces
Lead-Time	6 Days		30 Minutes

* Parts per Person per Hour

Figure 9.3   Sample before and after metrics.

There is no one single way to show or measure improvement. In reality, a combination of techniques is usually employed using metrics or symbols. Figure 9.3 is a simple example showing before and after metrics for a given work area. The burden of proof is always on the team or individual to establish that these gains are real and not simple Hawthorne-style improvements. Follow-up over time is of course required, and it is vital to know which action items caused this improvement result to occur.

In some cases, jobs can simply be easier as a result of Kaizen. The time required may not change substantially, but the quality of the output or the ease of the task may be increased. Symbols such as the Therbligs we used in analysis of current methods can also be used in creative ways to show before and after comparisons.

Pictures are often used as a way to show before and after conditions of a Kaizen event. We advocate the use of such imagery for the sake of clarity in presentation. A picture is often worth a thousand words. However, be extremely wary of using pictures to establish improvement levels. Just because something is different or better looking does not necessarily mean that improvement has truly occurred. Normally, it is a safe assumption but seek ways to combine quantitative measurements in conjunction with visual images for maximum effect in communicating Kaizen activity results.

Another important component of finalizing Kaizen involves updating standards and documents used at the work site or any other location. In Toyota, documents fell chiefly into the following areas: documents for training purposes, documents for work standards, and documents for standardized work. All sound similar but have separate purposes. Your organization probably has similarly functioning tools, so please draw appropriate analogies for comparison.

Documents for training are items such as job breakdown sheets and other operator-based work instructions that form a basis for instruction. Once Kaizen has taken place, these documents and all new major steps, key points, and reasons why need to be updated to reflect the latest standards of the job. Failure to accomplish this task will lead to eventual problems in training and errors in human execution as workers rotate or new people are trained for the job.

Another category of documents inside Toyota is referred to as work standards. These are technical documents that form the basic engineering, quality, and maintenance-related building blocks of the process. The items change less frequently but still are often affected by Kaizen activities conducted upon equipment. Sample items in this category might include quality control charts, quality check sheets, operation drawings, tooling drawings, setup sheets, gauging instructions, preventive maintenance routines, spare parts, or other such items. These also must be updated when affected by Kaizen activities to truly reflect the current state of the machine.

The final bit of documentation that is normally unique to Toyota is that of true standardized work-related documents. Kaizen almost always affects the operator-and-machine interface and as such the walk time, wait time, work time, and other such critical aspects of the job. For proper daily site management to occur, all forms of standardized work must be updated as part of the Kaizen follow-up and kept in concert with the actual operation of the machine.

In your company, you may have other types of documents and standards that need to be updated as well. Take the time to identify all such items affected by any Kaizen activity and make sure that standards are accurate and up to date.

The final aspect of Kaizen that we comment on is that of communicating results and sharing success. In Toyota, short (10- to 15-minute) presentations are always scheduled at the completion of Kaizen activities. The purpose of the presentation is to clarify the Kaizen story and make the team or individual be sure that he or she has conducted Kaizen in an appropriate manner consistent with the six main steps outlined in this workbook. Another purpose of the presentation is to recognize the effort of the individuals involved and to communicate the success to other parties.

Reporting Kaizen results should be kept simple in the spirit of this workbook and the philosophy of Kaizen. In general, only a few pages or charts need to be prepared. The presentation can take place with a flip chart, on a single A3-size piece of paper, or of course using a computer and projector. The Kaizen story should not be complex or confusing and the results should be easily understood by anyone attending the presentation. A sample outline for presentation is given in Figure 9.4.

9.3 Summary

In this final step of Kaizen, we emphasize the importance of verifying results in a simple yet quantifiable manner. Remember that change does not necessarily equal improvement, and activity does not equal accomplishment. Take a firm

Presentation Procedure

- Give a Brief Introduction to the Area
- Identify the Improvement Opportunity
- Explain Your Analysis Using:
    - Work Element Analysis
    - Motion Analysis
    - Time Study
    - Standardized Work
    - Process or Material Flow Analysis
    - Equipment Loss Analysis
    - Other Techniques Applied
- Explain the Ideas for Improvement
- Discuss the Implementation Plan
- Have a Plan for Evaluating the Method
- Hold Q & A Session

Figure 9.4   Kaizen presentation outline.

look at any improvement work conducted and always ask the hard questions of yourself and others: "How much have we improved?" "How do we know this to be true?" Be clear regarding how you are measuring improvement and make the standard or basis for comparison consistent and meaningful. When improvement has not occurred, the only honest and correct approach is to revisit analysis, generate new action items, and try out new implementation items. With such continual trial and error this Kaizen pattern was established and codified inside Toyota Motor Corporation over the past few decades.

9.4 Homework Exercise

In this chapter, the homework is simply to tell your Kaizen story. If you have an effective activity from the fairly recent past that is worth summarizing, prepare it in a sample Kaizen presentation report. If you have some work currently under way, use that as a basis for making a Kaizen presentation in the near future.

Chapter 10

Summary

10.1 Introduction

In closing, we offer some final thoughts on the main steps of Kaizen and the need for individual skills development with regard to this topic. In reality, there is no "one way" to do Kaizen in terms of steps, analysis methods, or timing (i.e., length and duration of activity). The main contents of this workbook regarding the steps of Kaizen were drafted with the intention of sharing the basic Toyota pattern as a starting point for training people to think about this topic and introducing some basic tools for analysis. Depending on your situation, you may have to alter and revise the contents and methods just as Toyota did several times during its improvement journey in the past few decades. Even today Toyota is emphasizing a back-to-basics movement. Here are some key points that we would like to reinforce regarding each step of Kaizen.

10.2 Step 1: Discover the Improvement Potential

The main point we want to stress concerning Step 1 is the need to open the eyes of the participant with regard to the nature of Kaizen and the tremendous opportunity that exists for potential improvement. There is no shortage of either waste or problems in the world. There is, however, a critical shortage of talented leaders in terms of driving improvement. The basic concepts outlined in this chapter are geared toward teaching people how to be more aware of the inherent opportunities that face them on a daily basis. In addition, this chapter is about how to begin to think about Kaizen. In other words, strive to properly identify and then eliminate waste. Do not simply fall for the trick of working harder or longer or throwing money at the problem. Waste identification, 5S concepts, production analysis boards, and other techniques are useful as simple ways to get started. The intent of Step 1 is to get people comfortable with the topic of Kaizen and to introduce the basic steps for moving forward.

10.3 Step 2: Analyze Current Methods

Simply observing an area of operations is often sufficient to highlight opportunity, generate ideas, and establish new goals. However, in some cases more horsepower is needed, and analytic tools are useful in breaking down operations into smaller pieces for study and improvement. The most fundamental tools available for analysis of current methods in Step 2 are process flowcharts, time study, motion analysis, and work element analysis. We recommend learning and adapting these as necessary for your situation. As in Toyota's case, you may be able to invent new combinations of these items, such as standardized work or material and information flow (value stream mapping) analysis charts. For that matter, the six major types of equipment losses may be of use as well.

Regardless of the type of analysis you perform, we strongly believe that the keys in virtually all cases are to be analytical, quantitative, and detailed. First, by *analytical* we mean selecting the proper organizing framework and tool for analysis. Organizing and characterizing the situation in an appropriate way is often half the battle. Second, by *quantitative* we mean crunching the numbers to show the exact extent to which different pieces of the puzzle add up to the whole. Items such as Pareto charts, for example, are invaluable guides when properly applied. Finally, by *specific* we mean being detailed in the sense of drilling down to the molecular level of detail. Toyota refers to this as being Genchi Genbutsu (actual location, actual object) oriented and possessing the skill of 5 Why inquiry. Developing these types of skill sets is critical in terms of importance for either problem solving or Kaizen. From our experience, there is no substitute for practice and hands-on learning in this area.

10.4 Step 3: Generate Original Ideas

In Step 3 of Kaizen, we move somewhat out of the analytical and reductionist world into the realm of synthesis and idea generation. Problem-solving activities often converge to a root cause and single solution for that situation. Kaizen is more flexible and encourages different ways of thinking about both the problem and the solution space. In the end, both techniques aim for improvement, so the difference is more conceptual than practical.

The magic ingredient of Kaizen is unleashing the inherent power we have as individuals to think creatively and in original ways about problems. For many of us, this is unfortunately not second nature. Fortunately, original thinking can be stimulated to a great extent, and teams working together can often achieve more than initially believed possible. There is some use for the checklists and guidelines supplied in Step 3. However, the best ideas will always come from inside and depend highly on your situation. Practice the art of brainstorming, and you will be surprised at the creative insights you can develop by yourself or in conjunction with others.

10.5 Step 4: Develop an Implementation Plan

The spirit of Kaizen is action oriented, and the case can always be made that the best plans are no plans, simply making change occur. In reality, even when no plans are committed to paper, a planning process is taking place inside our heads. Who will do what, where, how, by when, and why are fundamental questions that have to be answered whether the plans are thought through or written down on paper. When change is difficult and takes time, a plan is a good tool to keep people on track and focused on delivering action items as promised. We present some options in Step 4 as simple tools for helping to organize your thinking when needed.

10.6 Step 5: Implement the Plan

There is little of practical use that we can offer regarding the specifics of Kaizen implementation without knowing your particular situation, analysis, goals, or obstacles, for example. We do see common themes, however, in teams that either struggle with Kaizen or fail to achieve as much as anticipated. The top two problems are in the realm of communication and instruction. Kaizen involves change, and this by nature affects different parties in different ways. Err on the side of overcommunication in Kaizen and plan to obtain feedback early and often from affected parties. Conversely, ignoring people is never a good technique for obtaining cooperation or approval in a timely fashion.

Instruction is another area in which Kaizen implementation plans sometimes falter. Change drives a need for some form of training in most cases. If only the people participating in the Kaizen activity are aware of and participate in the change, then its impact will probably not be as great as expected. Take some time during implementation to make sure that all affected parties, including off shifts, those in material handling, and those in other support functions, are aware of the change and are provided proper job instruction.

10.7 Step 6: Evaluate the New Method

The last step of Kaizen involves verifying whether improvements have actually occurred and then standardizing the practices that have been improved. Making

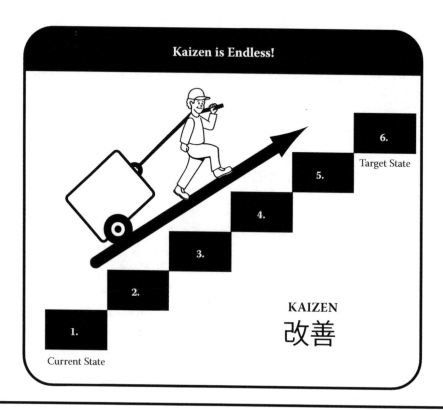

Figure 10.1   Kaizen is endless.

change is easy; making change for the better is more difficult but far more rewarding. The key point is to honestly measure the performance of the process before and after implementation. How you will measure it depends on what you are implementing and what your goals are for the activity. Kaizen without measurement and comparison is simply not Kaizen. Standardize what works and, more importantly, revisit anything that did not work and try again.

In summary, the process of Kaizen is similar to problem solving and the scientific method. There are subtle differences to each method, but all involve a methodical step-by-step thinking process. The goal in Kaizen, however, is not simply to follow the process; it is to execute the spirit of the steps and to generate a better process. This improvement challenge relies deeply on the skill of the individual and his or her ability to think clearly and probe deeply into the current state of affairs. The opportunity for Kaizen is endless (Figure 10.1), and the process is rewarding for those willing to undertake the challenge. We hope that this workbook is of use to you on your improvement journey.

Appendix 1: Ten Areas to Investigate for Operational Improvement

Here are 10 additional points to think about for improving basic operations of a typical process. The following list applies mainly to operations involving human work; however, they can be extrapolated to other cases as needed with some thought and effort.

1. Can we improve work motion?
2. Can we reduce variations in work element time?
3. Can we separate human work from machine work?
4. Can we revise standard work in process?
5. Can we reduce walking distance?
6. Can we better balance work among employees?
7. Can we improve the quality of the process?
8. Can we improve the reliability of the machine?
9. Can we store, locate, and transfer materials more efficiently?
10. What other points can we consider?

A1.1 Improvement Possibilities for Work Motion

First, we discuss improvements that can be made in motion related to work. Depending on the level of employee skill, large differences can occur in work motion. For operations to be performed efficiently, waste in work motion must be found and eliminated.

Tip 1: Learn to watch body posture for Kaizen points.
- Are employees working with bent or stretched backs?
- Does the angle of the body suddenly change?
- Can anything be done while walking?
- Does the direction of motion suddenly change?

Tip 2: Learn to watch hand movements for Kaizen points.
  – Are both hands moving efficiently? Is there any waiting time?
  – Are movements by the hands too large?
  – Are there any times when the hands just hold position?
  – Is the work too up/down or right to left?
  – Is the finish position okay? Is movement to the next step okay?
Tip 3: Learn to watch the eyesight line for Kaizen points?
  – Does the angle of the body suddenly change?
  – Are the hands spaced from tools and parts correctly?
  – Are parts, tools, gauges in position okay?
  – Is there any searching or groping motion to locate something?
Tip 4: Learn to watch the feet for Kaizen points.
  – Are there any movements away from the flow of the work?
  – Is there any work for which there is a long stoppage?

With the basics of Therblig analysis, you can break down any individual motions that might need further study for Kaizen. You can also obtain ideas for Kaizen by focusing on work elements and motions one by one. Try different techniques and see what works best for different cases.

Here are some questions to consider as well.

What will happen if we:
  – Change the angle of the part tray or shelf height?
  – Change the quantity of the part box or the size of the pallet?
  – Change the size or height of the chute or table?
  – Change the position or angle of tools, jigs, parts?
  – Change the order of the work sequence?
  – Change the start method?
  – Reduce the number of trips by carts and so on?
  – Use both hands?

All of these items can help you improve by cutting seconds from the work process and make the job easier and more efficient at the same time.

A1.2 Focus on Variations in Work Element Time

If you use time study methods or standardized work, there is often the problem of variation in work element time. The larger the variance, the more unstable the work element is and generally the more difficult.

Even when some jobs are standardized, there are often individual work elements that experience extreme time variation. Remember, if veteran workers experience fluctuations in work, this will result in bigger fluctuations for inexperienced workers. These types of problems need to be given high priority. The following are some items and tips about what to investigate for improving work with large time variations:

Is there poor workability?
- – Is the work hard to see, is there a lot of hand searching?
- – Is there a lot of adjustment?
- – Are lots of special knacks required?
- – Is there an element of muri/overburden in the job?
- – Are the tools easy to use?
- – Is the precision of tools, fixtures, and the like good?

Is the quality and shape of incoming parts good?
- – Are there any bad parts mixed in with the others?
- – Are the shape and precision of the part good?
- – Are parts easy to take from the box?
- – Is repair work occurring?

Are there any wrong or missing parts?
- – Are parts in their specified location?
- – Are the parts displayed clearly and easy to see?
- – Are work standards/standardized work created?
- – Are parts being delivered in small quantities?
- – Are kanban cards collected, sorted, and parts ordered?

Investigate carefully each of these points for improvement and consider ideas for each point needing Kaizen.

For problems of these types, the following items can be considered:

Improve the work method
- – Improve work requiring adjustment (eliminate need for adjustment)
- – Simplify work that requires skill or knack
- – Revise tools, parts, and the like

Stabilize quality
- – Improve the stability of part quality
- – Prevent wrong parts from getting in the box
- – Specify locations for parts and tools

Promote standardization
- – Create work standards and standardized work
- – Follow work site rules
- – Make infrequent jobs occur at periodic intervals

A1.3 Separate Human Work and Machine Work for Kaizen

Thinking about how to separate human work from machine work is an important concept for Kaizen. Humans need not be slaves to machines and be forced to monitor automated processes. No one watches the washing machine cycle at work, for example. When a machine is on automatic cycle, there are many instances when a person must stand and watch the machine. These represent opportunities for kaizen.

Look for improvement opportunities such as the following instances:

- Is anyone just watching a machine?
- After pushing a start button, does a person have to wait and watch for a while?
- Are people holding parts or adjusting the position of things?
- Are ejectors, chutes, and transfer devices functioning?

For instances of these types, some of the following can be applied:

- Eliminate unnecessary work or part trays
- Automate part feeds
- Automate part ejectors
- Set up chutes to feed the part to the next station
- Repair and maintain damaged part trays, jigs, and the like

A1.4 Revise Standard Work in Process

On manual lines, work in process points and work handoff areas are often set up without thorough consideration. For every unnecessary work in process point, an unnecessary series of motions will also occur.

- Is the standard work in process set correctly?
- Is the standard work sequence being followed?
- Is the work in process really necessary?
- If the work in process is necessary, can it be limited?

For these instances, you might try the following:

- Make it so that only the correct number of materials can be set down
- Clarify visually how many pieces of material should be in place

A1.5 Reduce Walking Distance

In a manual work line, the layout often gets set up with an equipment focus that leads to wasteful walking. Or, volumes and product mix change over time, and the current layout is no longer optimal from a walking point of view.

Focus on the following to reduce walking distance:

- Is the walk distance large between work points?
- Is the walk line between points straight?
- Is there any back-and-forth walking?

Think about the following ways to reduce walking:

- Change the layout where possible to be more accommodating
- Change the location of the parts pallet and part shelves
- Eliminate unnecessary obstructions

A1.6 Work Balance between Operations

For operations with multiple employees working together the work performed by each may vary, and the time required may not be balanced. This arrangement can often result in situations where one employee may be overburdened with respect to another. Or one or more operator might simply have a significant amount of idle time.

In these cases the best practice we can suggest is to use a standardized work combination chart and carefully map out the work content of each person in detail. Identify the main tasks and work elements for each person. Time study the work elements and create a comparison of between the operations in question. Balance the work in the most efficient manner possible by moving elements from one process to another where that is feasible.

Keep in mind however that you do not always necessarily want to balance operations evenly as odd as that may sound. The reason is that for maximum efficiency you want to balance work to takt time as fully as possible. In the case of an area or production line with seven people for example the last person might not have work that totals up to the level of takt time. In this case and others you would balance the work of the first six people and leave the seventh with whatever work remains. Over time you can look for ways to reduce this last operation through Kaizen. In the short term employees can rotate jobs to ensure some of opportunity and safety

A1.7 Quality Improvement Possibilities

There are endless opportunities to improve quality with respect to Kaizen. Indeed you may choose to simply purse traditional problem solving activities in these instances. However Kaizen is also an effective way to look at improving quality in many cases. Here are some generic points to keep in mind when working on quality problems in Kaizen activities.

- Establish simple red bins near the production line for storing defects for quick analysis
- Simplify the process related to the daily logging of scrap and rework types.
- Clarify the 80/20 rule for what are the dominant defect types.
- Establish clear rules for handling defective products
- Look for ways to prevent simpler defects like scratches, nicks, or dents
- Consider packaging changes for handling related problems
- Look for simple changes to the part design or processing method
- Maintain a clean work area to prevent contamination related problems
- Look for simple ways to improve process capability
- Ensure that tool changes and quality checks are done in a timely fashion
- Simplify gauging or inspection of parts

A1.8 Equipment Reliability Improvement Possibilities

Just as with quality there are endless equipment and maintenance improvement possibilities with respect o Kaizen. In the workbook we addressed the six major losses of equipment uptime on a process and those are a good place to review for improvement ideas. Here are some points to review when considering Kaizen in an equipment intensive area.

- Create a database of all major equipment breakdowns to study for repeat problems.
- Highlight the top five or so machines for improvement. This is often a good place to start.
- Interview operations personnel and maintenance to find minor stops that sometimes slip through the cracks of the tracking system.
- Practice observing equipment for a shift or two. This also highlights small equipment problems that tracking systems sometimes miss.
- Compare equipment breakdowns to the preventive maintenance items for the machine. Identify what is working and what is not.
- Look for simple visual inspection tasks that operators can perform without stopping the machine.
- Improve 5S situation in and around the machine.
- Try and make abnormal conditions obvious (e.g. low fluid levels, high temperature readings, etc.)
- Time study machine cycles and see if they are still operating as designed.
- Study any tool change or die change process and look for areas to simplify the process.
- Examine losses to scrap, rework, or start up yield losses and seek ways to minimize those losses.

In addition to these areas that focus on the equipment you can also look at the work of maintenance personnel and seek ways to improve this area as well. For example the following areas can be studied for improvement.

- Improve the collection of equipment breakdown data for analysis.
- Highlight the more repetitive maintenance calls and simplify these tasks.
- Perform a time study on several typical maintenance calls. Break down the repair time and see what wastes or difficulties occurred.
- Normally there is tremendous opportunity to improve machine drawings, electric circuit diagrams, spare parts lists, and general communication practices.
- Improve the spare parts storage area and minimize the time require to find tools and materials.
- Analyze the preventive maintenance schedule and its effectiveness.
- Make sure that all special tools required of maintenance are present.

Other angles for improvement exist as well within the realm of equipment and maintenance. Add to this list depending upon your own unique situation.

A1.9 Material Flow and Storage Improvement Possibilities

Most of the problems associated with material flow and storage can be highlighted drawing a good material and information flow analysis diagram. That technique is the normal starting point for identifying improvement angles in this area. The following suggestions may also give you some additional points to consider as well.

- Review containers and container sizes with respect to demand quantities.
- Review the number of times that material handling has to move and remove items.
- Look at minimizing storage locations and the number of associated moves.
- Clearly label and identify all material storage locations in terms of name, type, and quantity of items stored.
- Establish a plant wide addressing scheme that identifies material storage locations with ease.
- Color cold storage levels so that shortages are clearly visible.
- Consider dedicating paths that have high material handling needs to this function. Keep pedestrian walkways clear of these areas as much as possible.
- Quantify the amount of inventory that belongs in any dedicated storage area with respect to cycle, buffer, and safety stock.
- Establish rules and guidelines for the usage of buffer and safety stock.
- Review the nature of the material flow in the area and determine if replenishment, sequential, or mixed type pull systems are applicable.

- Review the disciple of material handling and determine of quantity based or time based systems are most appropriate for delivery.
- Study material handling routes and identify difficult points and areas for improvement.

A1.10 Other Improvement Areas

There are endless areas of improvement potential using Kaizen. In reality you will only be limited by your imagination and time available to spend on this topic. By design this simple workbook focuses on the Toyota method of studying basic production processes and beginning with improvements in these locations. Plenty of other areas for improvement exist however which are not covered in this introductory workbook. Any of the following areas might be fruitful to consider depending upon your circumstances.

- Look at energy costs and devise ways to lower any and all such costs that pertain to energy.
- Improve lighting and cleanliness of work areas.
- Study the oils, lubricants, other fluids, or other auxiliary materials used in the process and find suitable substitutes that might be used.
- Repair and prevent oil and air leaks.
- Improve safety concerns and areas with repetitive injuries.
- Study the overall delivery of material into a facility in terms of logistics efficiency.
- Identify work that might be performed more appropriately at a supplier or cases that might be brought in-house.
- Review basic aspects or assumptions pertaining to the design of the product. Looks for simple things that can be changed to make production easier.
- Evaluate simple changes to tooling, gauges, fixtures, clamping mechanisms or other devices to improve functionality.
- Study the production process as a whole and think about new processing methods that might make sense in the future.
- Improve the documentation surrounding the process such as standardized work, engineering drawings, or maintenance documentation, etc.

Appendix 2: Forms and Instructions

In this appendix, we provide a few more detailed points regarding five of the basic forms presented in this workbook. Analysis depends highly on your goal and the nature of the process. Forcing the wrong type of analysis will not result in useful insights. However, these five basic forms have a wide range of applicability. We suggest using them as a starting point when they match your needs or creating modified versions to suit your circumstances.

A2.1 Work Analysis Sheet

The work analysis sheet is the most basic and versatile tool that we have introduced in this workbook. You can easily modify this form to suit your needs, and it can easily be applied to just about any type of process. Here are some points of advice regarding this form.

Basic steps for work analysis (Figure A2.1):
1. Fill in any product- or process-related information.
2. Fill in the work elements for the process that you are analyzing.
3. Fill in any particular details for each step, such as distance, time, and ease.
4. Question why the step is necessary.
5. Question what the exact purpose of the step is.
6. Question where the step should be performed.
7. Question when the step should be performed.
8. Question who is best suited to do the work.
9. Question how the work element is best performed.
10. Jot down any improvement ideas that might occur to you.
11. Consider which details can possibly be eliminated.
12. Consider which details might be combined.
13. Consider which details might be rearranged.
14. Consider which details might be simplified.

Work Analysis Sheet

NO	Work Elements	Safety Distance Dimension Quality Ease	WHY	WHAT	WHERE	WHEN	WHO	HOW	Improvement Ideas	E	C	R	S

Figure A2.1   Work analysis sheet.

A2.2 Therblig Motion Analysis Form

The Therblig motion analysis sheet takes some practice but once mastered can provide many detailed insights pertaining to work. This technique was created for detailed motion analysis of manual operations. We suggest using this form of analysis only when it makes sense to do a detailed look into human motion. Be sure to familiarize yourself with the 18 basic Therblig symbols before using this form. Often, it makes sense to have more than one person look at the motions due to the detailed level of observation. For best results, include someone who has done the process before or is currently an operator of the process.

Basic steps for Therblig motion analysis (Figure A2.2):
1. Fill in the basic product- and process-related information.
2. Make a small sketch of the area you are studying for a reminder.
3. Observe the motions overall several times to obtain familiarity.
4. Write down the left-hand, right-hand, and eye-related motions one by one.

Therblig Sheet

PROCESS/ OPERATION NAME		KEYS			
PART NUMBER					
PART TIME					
INVESTIGATION	DAY-MONTH-YEAR	NAME			
REMARKS					

KAIZEN IDEAS	LEFT HAND			RIGHT HAND		KAIZEN IDEAS
	WORK ELEMENT	EXPLAIN	THERBLIG	EXPLAIN	WORK ELEMENT	

Figure A2.2 Therblig Motion Analysis Form

5. Include a brief word or two of explanation if required in the provided column.
6. Write down the work elements every few symbols for clarification.
7. Write down your improvement ideas for each small motion.

A2.3 Time Study Form

Time study is fairly complicated in many cases due to the nature of the subject being studied and the fact that the underlying operations are not well standardized. In many cases, it makes sense first to study and standardize the process in greater detail before attempting a time study. In other cases, of course, the act of time study may help you derive insights regarding what to further standardize for improvement. Correctly filling out a time study observation form assumes working knowledge of a stopwatch and the ability to identify work elements. For simplicity, we highlight some suggestions to follow for simplicity. Other methods can be used as well if you are more familiar with those methods.

Basic steps for time study analysis (Figure A2.3):
1. Write down the name of the product and process being studied.
2. Familiarize yourself with the entire cycle and work content.
3. Write down all the work elements you plan to time study.
4. Measure several total cycles first and record the times.
5. If the time variation is large, find out why and try to improve it.
6. If times are consistent enough, then proceed to the next level.

Time Observation Form

Time Observation

| | PROCESS | | | | | | | OBSERVER | | | | | | | DATE | |
| STEP # | Work Element | 1 | 2 | 3 | 4 | 5 | 6 | 7 | 8 | 9 | 10 | 11 | 12 | TASK TIME | REMARKS |

TIME FOR 1 CYCLE

LOWEST REPEATABLE CYCLE TIME

Figure A2.3   Time observation form.

7. Determine if you want to calculate elemental times for each element or groups of elements. For example, if there are 10 work elements, you might group 1–3, 4–6, 7 and 8, and 9 and 10 together for simplicity.

8. For each work element or group of elements, determine clear measurement start and stop points that are easily identified.

9. Measure and estimate your elemental times for the individual elements or groups of elements as needed.

10. Identify sources of variation or longer recorded times and probe into why they occur.

11. Take 5–10 cycles of data as needed to obtain accurate measurements.

12. Use most repeated times if you are looking to set an initial baseline for the process.

13. Use the shortest elemental times (and ask why that time was possible) if you are looking for improvement potential.

14. Write down any comments or ideas you might have in the remarks column.

A2.4 Standardized Work Chart

Standardized work is a specific document in Toyota that takes some time to master. Covering all the details of standardized work would take as long as this entire workbook on Kaizen. For a taste of what the original standardized work course inside Toyota is like, you can visit the Art of Lean Web site (http://www.artoflean.com/) and examine the document section. In the document section is a sample of the five-day standardized work training course from Toyota. Completing a standardized work chart often also involves a process capacity sheet and a standardized work combination table. For simplicity, we outline just the basics of completing the standardized work chart in this appendix.

Basic steps for time study analysis (Figure A2.4):

1. Fill in the required information for product and process and the like.

2. Clarify the takt time for the job in question. Jobs that do not have a clear takt time might be better served by some other form and type of analysis.

3. Identify the major steps of the job possible under the limit of takt time. (Note: The standardized work combination table helps in this step.)

4. Make a sketch of the area and cyclical work being performed.

5. Time study the major steps for the process being studied. (Note: Calculate the amount of manual time, machine time, wait time, and walk time. The standardized work combination table and process capacity forms help with this step.)

Standardized Work Chart

								SAFETY	SWIP	QUALITY
Standard Work Chart						PLANT:		PRODUCT:		
						AREA:		Op. ___ of ___		
DATE:	BY:		APPROVED BY:			PROCESS:		Pg. ___ of ___		
						SHIFTS:				
						VOLUME:				
NUMBER	MAJOR STEPS	Man. Time	Auto Time	Wait Time	Walk Time	WORKING SEQUENCE / WALKING / RETURN TO START		✚	◐	◇QC◇

Figure A2.4 Standardized work chart.

6. Identify any major safety or quality points in the operation and insert the proper icon in the appropriate location.
7. Identify the correct number of standard work in process pieces of inventory.
8. Identify main areas for improvement given the current takt time and operator cycle time.
9. Pilot improvement ideas and determine effectiveness.

A2.5 Setup Reduction Analysis Form

Setup reduction analysis is a great technique to apply, especially when a change-over in production occurs. This technique can apply to other instances as well, depending on your circumstances. Skill in setup reduction analysis assumes basic familiarity with the process studied, identifying work elements, conducting a time study of work elements, and problem solving. We outline some of the common steps involved in this process.

Set Up Reduction Worksheet

NUMBER	Main Setup Work Elements	TIME STUDY			CATEGORY		PROBLEM POINT	COUNTERMEASURE
		START	END	TOTAL	INT.	EXT.		
1.								
2.								
3.								
4.								
5.								
6.								
7.								
8.								
9.								
10.								
11.								

LINE NAME

PART NAME

PROCESS NAME   APPROVED BY: PART NUMBER

Setup Improvement Analysis

Figure A2.5   Setup reduction analysis form.

Basic steps for setup reduction analysis (Figure A2.5):
1. Fill in the basic product- and process-related information for the area in question.
2. Familiarize yourself with the overall work.
3. Film the process if possible for study purposes.
4. Identify the major steps or work elements in the process.
5. Measure the time required for each step in the setup and the overall time required.
6. Determine whether the step should be done in internal time when the machine is stopped or in external time when the machine is still running.
7. For each step, isolate the main problem points and identify countermeasures.
8. Focus on the steps that are the most time consuming or most difficult to conduct for bigger breakthroughs.
9. Alternatively, set a goal for improvement (e.g., reduce by 20 minutes) and pick areas to work on that are likely to yield the required improvements.
10. Be sure that all tools and information are on hand when needed.
11. Move as much internal work to external work as possible. Conduct external work before the machine shuts down for the changeover.

12. Study and reduce the external times in detail.
13. Study and reduce the internal times in detail.
14. Eliminate need for adjustment by better alignment or setup method.
15. Eliminate or reduce fastening time due to turning wrenches. Use clamps if possible.
16. Use quick-disconnect supply lines for any needed fluid hoses.
17. Pilot the improvement ideas one by one and determine effectiveness.

Index

About the Author

Isao Kato is a retired manager of Toyota Motor Corporation in Japan. Kato spent 33 years at Toyota in a variety of positions before retiring to start his own consulting company in 1993. During his tenure, Kato spent time in machinery maintenance, personnel development, education, and training, as well as guiding suppliers during a stint in Toyota's operations management consulting division. In particular, Kato provided extensive support for Toyota's overseas facilities in training and education in the Toyota Production System.

Internally at Toyota, Kato specialized in codifying materials pertaining to a variety of topics associated with the Toyota Production System. Kato was a master trainer for Toyota in the Training Within Industry (TWI) materials including Job Instruction, Job Relations, and Job Methods. Under the direction of Taiichi Ohno, Kato also edited the first Toyota Production System (TPS) manual in the early 1970s and established a variety of TPS training courses. Notably, Kato developed the majority of the content that comprised the underpinnings for Standardized Work and Kaizen training courses for several decades.

Since retiring from Toyota, Kato has worked with companies in Japan, South Korea, Great Britain, Canada, Australia, China, and Italy. Kato resides in Toyota City in Aichi Prefecture in Japan.

Art Smalley is an expert specializing in the area of world class methods for operational improvement and has served numerous major companies around the world. Art was one of the first American's to work for Toyota Motor Corporation in Japan first studying at different universities in Japan and then learning TPS manufacturing principles in the Kamigo engine plant where Taiichi Ohno was the founding plant manager. During his stay in Toyota Art played an instrumental role in the development and transfer of both precision equipment and TPS methods to Toyota's overseas engine plants. Art was responsible for multiple facets of engine production, equipment planning, maintenance, and tooling.

After a decade in Japan, Art returned to the United States and served as Director of Lean Manufacturing for Donnelly Corporation a $1 billion tier one automotive parts supplier. Art subsequently joined the international management consulting firm of McKinsey & Company and was one of the firm's leading experts in the area of lean manufacturing for a period of four years. During this time he counseled numerous Fortune 500 clients on operational matters involving lean implementation and lead specific cost, quality, and delivery improvement projects.

In 2003 Art launched his own company Art of Lean, Inc. and now divides his time serving a diverse base of manufacturing clients such as Parker Hannifin, Delphi, Schlumberger, Gillette, Sandia National Laboratories and many other companies in areas of operational performance improvement.

In addition Art serves as senior faculty member and periodic advisor to the Lean Enterprise Institute and its global affiliates delivering lectures to leading manufacturing executives around the world. Through the institute Art published the definitive workbook guide on implementing basic pull production methods entitled "Creating Level Pull" which was awarded a Shingo Prize for distinguished contribution to manufacturing knowledge in 2005. In 2006 Art was inducted as a lifetime member of the Shingo Prize for Excellence in recognition of his contributions to manufacturing. In 2008 Art also co-authored the Shingo Prize award winning book entitled "A3 Thinking" with his friend and colleague Professor Durward K. Sobek.

In his spare time Art enjoys woodworking, playing golf, and reading books.